# Treachery

## by

## Andrew M. Patterson

In commemoration of Mehmet Savas Ozcan and the heroic officers and police of Turkey, whose *Peace in Turkey and Peace in the World* of the July 15, 2016 was the failed coup attempt to oust Turkish President Tayyip Erdogan whose son, Bilal was buying ISIS stolen oil and selling that oil at $8 a barrel to Israeli and Turkish oil companies with permits from the Ministry of Energy headed by Bilal's brother-in-law/ ISIS then had money and bought military hardware, trucks, munitions, and supplies it needed to continue their wave of horror across northern Syria and Iraq. ISIS used some money to pay volunteers. But as Chechens from Russia saw ISIS as false Islam and gave Vladimir Putin records of who was paid and how much. ISIS traced this, information back to the Chechens and they were executed. *Treachery* exposes treason of people in the CIA, FBI, Pentagon, acting on behalf of Israel's Mossad and the connection to Brazilian Jews.

ISBN    13: 978 1530736621
ISBN    10: 1530736625
Re-edited            September 21, 2017
Re-edited again   March 8, 2018

# Cover

Upper left corner Mossad Mike Harrari who admitted he planned 9/11. Mid photo: flames in the North Tower. Upper right, Larry Silverstein who built the WTC in 1972, and owned Building Seven he had pulled. It had the back-up files for the 2.3 trillion dollars Rabbi Dov Zakheim stole from the Pentagon budget. Bottom photo: shows the crater in bedrock made by a mini-nuke bomb place on the basement floor that vaporized the bedrock and melted the fireproof steel at an estimated 6,500 C. the same temperature as the liquid iron of the upper mantle of the Earth's core as measured by French scientists in 2015. The author went to the atomic physics department at the U of Illinois and was escorted from a classroom building saying I had no right to be there. The assistant head of atomic physics said he was too busy to check the figure used by the author. *The Fix was In* and if any employee of a university, media, architectural/ engineering firm said anything different than the NIST report they'd lose their job just like the man at a book fair in Berlin who listened to my description of my book and said he thought his boss would be interested. When he returned an hour later he said, "Please leave I don't want to lose my job." . ...

A poll taken in the US in 2014 had 38% of American stating they weren't sure the official 9/11 report was truthful. 10% were sure the official NIST ort was covering up 9/11, 12% had no opinion while 40% believed the NIST report that suicidal Muslims from Saudi Arabia came to America, learned how to fly, hijacked planes and flew them into buildings.

In 2003, half of the Americans polled believed Al Qaeda did 9/11 and 80% believed Saddam Hussein was involved. But in New York City, forty nine percent did not believe the official report while 50% of Americans outside New York believed the American Government NIST report was true and accurate and anyone who didn't was anti-American or wrong.

In 2015, more than half the Americans believed Lee Harvey Oswald was the lone assassin of Presiden00 t Kennedy and many were told President Obama was born in Africa not Hawaii. There was an article in *Saturday Evening Post* in 1962 that said many people of mixed marriage came to Hawaii to avoid prejudice and a Hawaiian birth certificate was found

The author has written seven encyclopedia articles, five economic feasibility studies, worked on Latin America economic statistics for seven years with the Inter-American Statistical Institute, and worked with the International Finance Corporation, where he was handed a book of all the banks in the world with names of all the officials, the net worth, the securities invested in, countries they invested in, addresses, and telephones, just everything you could think was needed.. The book was the size of an annual *New York Times Index*. But no one can buy this book. It was handed to the author by his superior in IFC to assess which banks would be interested in investing in Latin American shares. The mission to Brazil was headed by Polish Jew who smuggled weapons to Jews in the Warsaw Ghetto in 1943 to make the Warsaw Ghetto uprising.

The author wrote about the Latin American stock Markets and the World Bank president added strengthening stock markets to the International Finance Corporation. Sent t Brazil, the author spoke in a Rio conference on how to lower the cost of money was well received but the missions failed because the mission leader was a fireman across from the Warsaw Ghetto and smuggled guns to the Jews in the Ghetto through a tunnel that created the Ghetto uprising.

## Table of Contents

# Treachery

Paul Warburg wrote the 1913 Federal Act that created the Federal Reserve that financed WW I for Britain, France, Russia, and all the Allies. The Secret Treaty of London in 1914 had Britain, France, Russia, Serbia, and Greece. A plan to assassinate Ferdinand Hapsburg in Sarajevo was to draw Austria into war and Germany would defend Austria, and the Sultan of Turkey would come to defend Bosnian Muslims against Serbs that launched its Black Hand against Bosnians in 1912 to get an outlet to the Sea. Turkey's Sultan declined as he said he could not dishonor its British Protectorate. But the main purpose of the war was to get Turkey into the war, destroy the Ottomn Empire and get Palestine as a Homeland for the Jews. That was done by Paul Warburg who wrote the 1913 Federal Reserve Act, invested in the bank along with seven European Banks and two American banks (J.P. Morgan and John D. Rockefeller, all of whom were Jews.

The Federal Reserve financed WW I for the Allie with American money. The war made 21,000 new millionaire and the Federal Reserve then withheld dollars from the banks breaking the banking system and the new millionaires became the robber barons who continued the financial rape of America. The collapse of Wall Street dried up the source for capital and purchasing power of the markets.duke Ferdinand in Sarajevo by the tubercular infected Princip, Britain, France and Russia say he cannot be guaranteed a fair trial by the Austrians and declare war. Kaiser Wilhelm supports Austria and the French invade. Paul Warburg whose bank is of the Federal Reserve,

Paul Warburg went to Kaiser Wilhelm to offer his help and the Kaiser is surprised that Warburg has more intelligence than his own German intelligence. The Kaiser exempts all Jews from military Service to get their cooperation for the War effort.

6

Warburg became so powerful that his bald head wasw the prominent feature in the Comic Strip. *Little Orphan Annie in* which Daddy Warbucks is compassionate to adopt an orphan whose father was lost in the Great War. Daddy Warbucks was always fighting against the forces of evil, but Warburg could be said to have an anti-social personality disorder as he had no remorse or guilt in starting wars to satisfy the Rothschild desire to create a Homeland for the Jews in Palestine. Nathan Rothschild tried to buy Palestine in 1832 from the Sultan who refused as the Sultan said if the Jews owned Palestine, they'd remove Christians and Muslins. The Sultan said Palestine was a Holy Land for Jews, Christians and Muslims and Turkey would keep it that way (i.e. *The Age of Disinformation)*. So Warburg tricks Kaiser Wilhelm into tricking Turkey into the war into WW I to pay off the participation of Britain, France, and Russia to go to war, they will get pieces of the German and Ottoman Empires. Warburg sands in agents to get 600,000 Jews conscripted into defect and gets Kaiser Wilhelm to send Lenin to Helsinki and by train into St. Petersburg to make the Communist Revolution.

Warburg tells the Kaiser to get Turkey into the war as Turkey is close to Baku oil the German navy needs and to trick Turkey into the war by giving Turkey two German cruisers. After all the crew given red fezzest, Turkish flags flying, and 20 Turkish officer to train in running the ship were on board, the German captains sailed  to Russian Black Sea ports, bombarded them and Russia, France, and Britain declare war on Turkey that professes its innocence..

The Allies knew in advance of Warburg's plan, and Winston Churchill plans the invasion of Gallipolis using Australian, New Zealand, and Indian troops in the biggest battle of WW I.

Next, Warburg tells Kaiser Wilhelm he can get 600,000 Russian Jewish conscripts to desert and the German Army pours through the holes in the line in six months Russia loses six million men.  Czar Nicholas abdicates but Prince Kerensky keeps Russia in the war.  So Warburg tells Wilhelm to bring Lenin from Stockholm to Helsinki and then by train to St. Petersburg.  Lenin makes speeches but no one pays any attention to him.

So Jacob Shift of the Federal Reserve calls Trostsky from his job on a New York newspaper to come to Washington and gives him twenty million in gold coins to carry to Russia and buy the Russian Revolution.

Lenin takes over and sets about consolidating his power.  Even though Czar Nicholas has abdicated and he or his son could never be Czar again, the British try to rescue him and his family as they are cousins to the Royal family of England and Lenin orders the Romanovs are butchered and the world is horrified.

Lenin becomes promiscuous and contracts syphilis that goes to his brain.  He is so unstable the Communist Party keeps him isolated from the public.  Itt is just a matters of time so Stalin and Trotsky began to vie for the leadership and Stalin wins out.

In Germany, the Communists began to demonstrate for a republic and battles broke out between those loyal to the Kaiser and the Communists.  To facilitate the Communist takeover, Warburg goes to Kaiser to tell him to abdicate and flee to the Netherland for sanctuary or the Communists wll kill him and his family like they did his cousin, Czar Nicholas. Wilhelm abdicates, flees to Holland, and then war is over.  The war could have continued as the German line was impregnable.

The proof is that when a British general claims he can break the German line with human wave tactics like Mustafa Kemal Ataturk used to force the British forces back to the beaches at Gallipolis. The general is given 500,000 Indians of the 950,00 volunteers for the British Army to fight in Europe.

When the British asked Indian leaders for one million men to fight in Europe, they replied, "Independence now and you may have them.. The British reply, "Not now, after the war."  Indian leaders say, "Help yourself."  But India has a bumper rice crop.

Rice is plentiful and cheap and no one volunteers. So the Rothschild buy up all the rice and ship it out of India and 950,000 starving men volunteer but fifty million men, women, and Children starve to death in India.

For three years, American rice farmers cannot sell one grain of rice and are told, "No one wants rice any more. So the author's great uncle and all rice farmers walk away from their very profitable rice farms and the Rothschild have a second huge *huge collective type used as collective farms in Russia and future Israel* and it only cost fifty million lives.

The Rothschild banking family financed the building of all railroads in the world with bearer bonds that can be cashed in another country. The number guarantees anonymity while the name, Rothschild is a guarantee of no default.

British Baron Paul Julius Freiherr von Reuters, was born in Germany but immigrates to Britain. Made a baron, he gets the contracts to build the telegraph lines to serve all railroads being built around the world as all were financed by Rothschild Bearer Bonds. Those telegraph lines became a very important source of intelligence with messages sent by governments and businesses were decoded by Reuters and analyzed for importance. Reuter's de-coding staff infiltrates British intelligence as well as intelligence agencies of other nations. Reuters developed into Reuters News Agency used by BBC and New York Times, associated Press and Jews from countries around the world had no problem in getting permission to work in other countries as translators to AP, CIA, MI6, to manipulate foreign policies of nations to favor Israel..

Communists in the Weimar Republic government took over the treasury of Germany. German Jews began buying property with ten or twenty percent down and the banks supplying the rest. When all the banks were mortgaged to the hilt, the Communists in the treasury began printing marks so fast that they became worthless. If a new pair of shoes cost one million marks the price

9

would be one million one hundred thousand or more tomorrow. Unemployment was estimated as fifty percent, and the National Socialist Party (Nazi) gained in popularity and filled the majority of seats in Germany's Parliament. Hitler became Chancellor on January 30, 1933.

The Communists were kicked out of government jobs and some put in Concentration Camps. Germany was taken off the gold standard and the old mark was scrapped and 500 Reichmarks were given to every man and woman in Germany and in a month, factories were producing goods for export.

New York lawyer, Samuel Untermeyer called for a meeting of Jews from all over the world to meet at the Bilderberger Hotel in Holland in June 1933. Winston Churchill was instrumental in starting was against Germany in WW I by saying "Germany's modern navy was a threat to Britain's domination of the sea in the continuation of the se." In 1942, he replace Nevil Chamberlain as Prime Minister and used the League of Nations to declare war on Germany after Hitler invaded Czechoslovakia over the massacre of three thousands of ethnic Germans by Betar terrorists of Riga and invaded source Poland as the Betar terrorists massacred seven thousand ethnic Germans in Poland. Things went badly for the giant army of the League of Nations as the superior German tanks, planes, and weapons ran over the defenses of the League of Nations and the British retreated to the beaches of Dunkirk to await rescue by Britain called on all boat owners to sail for Dunkirk to rescue 100,000 British soldiers. It was a herois effort by thousands of private boat owners but England was defenseless with all its old tanks and trucks left behind. The Royal British Air Force was the only defense Britain had and Franklin Delano Roosevelt had to save Britain because that held the Mandate for Palestine as a *Homeland for the Jews* and Sir Stafford Cripps was sent to Stalin to get him to attack Germany from the east. Stalin wanted the three Baltic nations. Eastern Europe, and the Balkans and he got what he wanted. But if the Russian Generals knew he was planning to attack Germany, they would have thrown him out and brought back Trotsky. So Trotsky snet a killer to Mexico to kill Trotsky who was killed two months later and sentenced to twenty years in prison. A Mexican prostitute was paid to serve as a sexual partner every week and she was paid from the United States. Mexico was never able to trace that flow of cash form the US.

In August 1961, the people of the author's apartment house in Washington were sitting outside watching America's space balloon going around the Earth. Someone asked, the author "Is America ahead or behind Russia in the race for space?" When the author replied that according to what he had read in the newspapers, America was behind. One surly man said, *"Well, if you like Russia so much why don't live there."*

That is an example *Chauvinism* that Black Op planners use to manipulate. Cord Meyer of the CIA was in charge of CIA's *Mockingbird* that was to feed lies to the news media that the crime of terrorism will be deflected and direct on others..

John F. Kennedy wanted to eliminate the power of secret societies of which the Illuminati is the most infamous and is connected Satan as well as Communism. It has focused on recruiting from the highest educated of teachers, university professors. But among such an intelligent group which one is to lead? Code Words like *bifurcation* were used so when a speaker said that word it meant for the others to *sit up,* pay attention, and follow his lead.

On September 11, 2001 thousands of American traitors working in the government to protect America were complicit in the murder of 3,000 men and women. Explosions took away stairways, exit doors were chained and the heat from mini-nukes in the basement traveled up 338 elevator shafts to cremate alive or vaporize those in the elevator cars. Those on the roof of the North Tower jumped as their feet were burning.

An example of a big lie that the most people accept as the truth is that Americans *invented the atomic bomb*. Americans did create the cyclotron at Batavia, Illinois outside Chicago that can split atoms but it won't produce an atom bomb. The secret of making an atom bomb is in the centrifuge.

Americans physicists working on the Manhattan Project had no idea that uranium ore had three isotopes: U234, U235, and U238. U235 is the most radioactive and the *trigger*. If any of the other isotope is mixed with U235, it will not work. U235 is about 1% of *yellowcake*. U238 is about 98% and can be made into U239, Plutonium which is highly fissionable.

In WW II, German scientists developed a centrifuge that spun at ***1,500 times second or 90,000 times a minute.*** The blades in the center had to be strong and balanced to withstand such speeds. The top bearing was a magnetic bearing and the bottom was a pivot. Hexafloric acid poured on uranium *yellowcake* to make hexaflouric uranium gas to put into the centrifuge. U238 was heavier and went to the outside and down. U235 was lighter and went up to the next centrifuge. After passing through 1,000 centrifuges pure U235 was obtained. The gas was then treated with calcium to make a uranium salt and the salt was then turned into metal U235. U238, (depleted uranium) treated with deuterium to make U239 (*plutonium*) the explosive element, and U235 was shaped into inverted hollow bell-shaped cone for the trigger. Ignited by an electrical charge from a battery produced 5,500 C. in the center of the cone shot forward into the *plutonium* to start the fission. If any other isotope was in the U235 it wouldn't ignite.

SS General Karl Wolff, head of German intelligence in Italy and SS General Rainer Von Gehlen, head of German intelligence in Eastern Europe, began negotiating surrender to the British and Americans in February 1945. In March the two atom bomb tests were successful. A German general tells Hitler with two bombs he could drive the Russians back to Moscow.

Hitler is elated and call Mussolini where the war is going badly. Mussolini says he wants to be there in Berlin to help and picks 18 top generals and tells SS Generl Karl Wolff he will fly into Berlin to be with Hitler. Wolff convinces Mussolini to go by car *as the Allies control the air over Italy.*

So a German anti-aircraft battery is added to protect them. Benito and Clara Petacci in the front of the motorcade in Clara's 1939 Alfa Romeo that Benito Mussolini gave her. On getting into Switzerland, Germany's top test pilot Hannah Reitsch would fly them into Berlin in a small plane landing on a street.

But the routing of Mussolini's motorcade was sent to the British and they sent it to Serbs brought into Italy by the German Army as slave labor. They had escaped and formed their own *partisan group* by killing German soldiers to get weapons. Every time they killed a German soldier, the Germans kill 50 Italian men are shot. The British liked that as it made Italians angry at the Germans. All Italians were innocent of killing German soldiers.

April 13, 1945, Roosevelt dies of a heart attack. April 20, Hitler's birthday and everything looks good for the plan to push the Russians back to Moscow with two atom bombs. But as Count Folke Bernadotte delivers the monthly report on the International Red Cross inspections of concentration camps and prison inspections, SS General Heinrich Himmler, head of German intelligence Himmler tells Bernadotte that Hitler will soon be dead. Bernadotte tells Hitler who is furious and orders Himmler shot on sight and Himmler's personal representative was shot. He was angry with betrayal of German intelligence and the disappearance of the two atom bombs, as well..

April 27, 1945, Mussolini's motorcade runs into the Serb ambush. The Italians surrender without firing a shot. The Serbs call London to ask what to do. Churchill has said *who has knowledge of how to make the bomb will rule the world once the war is over.* Everyone must be killed including Clara Petacci and the German anti-aircraft battery assigned to protect the motorcade from any possible attack..

The morning of April 28, an Italian speaking partisan tells everyone they will be freed. Just hand over any weapons they may have kept. They are then loaded into a van, driven out of sight of the village ordered out of the van and killed. Clara jumps in front of Mussolini to protect him and shot first. Mussolini bares his chest and say "shoot me in the heart." The Serbs drive the van to a Milano Esso station, hang the bodies up and begin shooting, beating, and spitting on Mussolini's body.

April 29, 1945, Karl Wolff and Rainer von Gehlen surrender to the Americans. The two atom bombs disappear and Hitler is devastated. The next morning, Hitler and Eva Braun marry and then commit suicide. Otto Skorzeny who saved Mussolini in July 1943 from the top of Gran Sasso is arrested, tried for war crimes but charges are dismissed. Two SS officers in US Army uniforms come with documents that he is to be turned over to them and Skorzeny, is taken to Spain where he is flown to Argentina and assigned to be Eva Peron's bodyguard. Karl Wolff is imprisoned with top Nazis at the Nuremburg trials and used as a snitch by Allied prosecutors.

Dr. Rainer Karlsch, wrote *Hitler's Bomb* and was denounced by other German historians. To be a celebrity or an *expert,* you follow the official line, for perks, privileges, promotions, top jobs, academic awards, and fat pensions.

In 1934, 33rd Degree Mason, President Franklin Delano Roosevelt, ordered the symbol of the pyramid topped by the *all-seeing eye* put on the back of the US dollar bill so everybody that has a dollar bill carries this Illuminati symbol with them. Interesting. Illegal drugs go to finance the New World Order and enslavement of mankind and extermination billions of people as the NWO says the planet's resources cannot satisfy the needs of all the world's population.

Palestinians have been occupied by an enemy that invaded them, Israeli, but its Shekel was the worst currency in the world so Saddam Hussein supported them. When no one was helping Bosnia, Saddam offered WMDs to Izetbegovic *to stop the Serb atrocities* and Izetbegovic said Bosnia had developed WMDs His empty threat led to *Saddam* and Susan Lindauer, a pacifist is made a CIA *asset*, and meet with Iraqis to destroy their WMDs. She said they were doing all she asked and anxious to avoid war but she was dismissed in 2001 five days after George W. Bush is inaugurated.

When Susan went public to say the Iraqis was doing everything to eliminate WMDs, she was imprisoned under the Patriot Act and held for five years. During this time the court refused to try her saying she was not mentally fit. The *government* offered to put her on Haldol which has impaired the minds of many who had been treated with it but the court said no and freed her.

The theft of 2.3 trillion dollars by Israeli-American Pentagon Comptroller, Rabbi Dov Zakheim, was covered up by 9/11. Dov's grandfather was Bolshevik revolutionary married to a relative of Karl Marx. Dov had top level security clearance as comptroller when he should never been allowed inside the Pentagon. Given control of 3 trillion dollars on May 4, 2001, three months later 2.3 trillion was missing. Dov's account books were destroyed by a Russian missile that penetrated the steel reinforced concrete wall of the Pentagon and the backup files in Building 7 were destroyed by Larry Silverstein who ordered it *pulled.* As the demolition charges were planted before 9/11, Larry was complicit in the 9/11.

92% off all Jews are ethnically Khazars (Ashkenazi), without any historical connection to Palestine. They created a kingdom of over one million square miles at the top of the Caspian called Khazaria and it held for 1,000 years conquering 24 tribes around them. The source of Khazar wealth came from

15

robbing merchant caravans going between Europe and China. The Khazars use the Hebrew alphabet to write their language as for 1,000 years they couldnot read of write. In converting to Judaism, they used the Hebrew alphabet to write Yiddish which only they can read. The first Communist newspaper in America was published by Emma Goldman in Chicago in Yiddish and the $600,000 to finance the paper is from *Victims of the Chicago Fire* collected in glass jars placed on counters in Midwest stores.

Immediately after 9/11, Israel attacked Gaza which was only 35% of the territory it was given after what the UN designated as Gaza. At stake was the huge gas field in the territorial waters of Gaza. Israel wanted to destroy Hamas, the legal government of Gaza and annex Gaza, expand its borders to include all oil fields of the Middle East into what will be Greater Israel, which was the goal of the Betar Terrorists. They formed in Riga, Latvia in 1923 and killed thousands of ethnic Germans in Czechoslovakia and Poland to get Hitler to invade those countries to protect the ethnic Germans.

President Dwight Eisenhower in 1961 warned *about the Military Industrial Complex* (MIC) corporations that the Pentagon continually awarded contracts for weapons, fuel, uniforms, food, and everything needed to wage war. It is these companies that back and support the Bush Administration's 9/11, the wars in Iraq and Afghanistan along with Project for a New American Century (PNAC) that wants *continuous* war to keep America alert. Its members are pro-Zionists who hold top-pay jobs in the Pentagon, CIA, State Department, and Justice Department and have called for a *New Pearl Harbor* and gave false information, assisted in the cover-up, and they are given top jobs from one agency to another and MICs with Homeland Security contracts, and think tank reports looking for terrorists.

President George W. Bush used 9/11 to invade Iraq and Afghanistan that has killed millions of their people. He issued a gag order suppressing *all evidence of Israel's involvement in 9/11* then gave orders to kidnap, detain, and torture Muslim detainees he knew were innocent. The White House legal counsel, Alberto Gonzales advised that all detainees must be kept out of the US to deprive them of the protection of the Constitution. Gonzales said, "The Geneva Convention on treatment of prisoners is *outdated* and *unrelated* to realities of today," He now teaches international law at Harvard.

President Bush pushed for the Patriot Act that took away the freedoms guaranteed in the Constitution. Senate Majority Leader Tom Daschle opposed that and his secretary was poisoned upon opening a letter to Tom three weeks after 9/11. The envelope contained a trillion spores of silicon coated anthrax. The anthrax was traced to the biological weapons laboratories at Dugway, Utah and Fort Detrick, Maryland that President Nixon had ordered closed in 1972 but is was kept open and made synthetic viruses by injecting RNA from animal diseases into the DNA of human viruses to make new viruses to be used in biological warfare.

# Introduction

Aldous Huxley said, *"In times of universal deceit, the Truth will be Revolutionary."* But these traitors know they are protected and will be rewarded.

A Black Ops operation is to put blame on someone else to turn a nation people to look to their government for protection On 9/11, Cord Meyer of CIA *Mockingbird*, got the media was to use *fanatic Muslims* whenever saying 9/11,

In 1962, the Pentagon planned a series of bombings in the Baltimore area to blame on Fidel Castro and start a war with Cuba. When *Baltimore Sun* reporters learned about the plot, they published it and when President Kennedy learned about he made sure the plot was cancelled. That I the difference between President Kennedy and Mr. George W. Bush.

John Lear, a CIA covert operations pilot for 34 years, has revealed that after President John F. Kennedy ordered Israeli Prime Minister David ben Gurion to put Israel's nuclear weapons under UN Inspection to conform with international treaties, Ben Gurion responded with, "Kill Kennedy."

John Rich, born in the Philippines into a wealthy Jewish family and James Angleton of the CIA, known for submitting false reports, were given the order to murder John F. Kennedy. They manipulated the Top Brass of the Pentagon angry because JFK was cutting their budget to make edonomic development all over the world. Then those who wanted war with Russia and invade Cuba. Then those in the Secret Service, FBI, and military industrial complex who collaborated in killing President Kennedy for personal gain. JFK's successful negotiation with the Russians the year before for the removal of Russian missile and nuclear warheads. Did he know about the Russian Czar bomb 1,570 more energy than the two German bombs used on Hiroshima and Nagasaki, I think he did.

Rich and Angleton looked for those in the Secret Service who were condescending of Jack his affairs and Jackie who was guest of Aristotle Oneness, Greek Shipping tycoon. Lyndon Johnson was told that JFK planned to discard him as his Vice President in the next election angering LBJ. He could have blown the whistle and foiled the plot but didn't as Ladybird wanted to live in the White and replace Jackie who she despised.

Santos Tafficante and Meyer Lansky, head of the Jewish Mob, invested in Casinos, hotels, sex shows, and whorehouses in Fulgencio Batisita's Cuba. Traffiante called Jack Ruby from Florida and told him to kill Lee Harvey Oswald to keep him from testifying in court, and Ruby did it the next day.

John Roselli of the Mafia trained snipers that William Harvey of CIA's assassination unit, and Sam Giancano of the Chicago Mafia was murdered for talking.

JFK planned to do away with the Federal Reserve which has kept America in perpetually in debt as is the World Bank and IMF by Zionist Jews, destroy the CIA, negotiate an end to the Cold War, peace with Cuba, peace between China and Taiwan, put nuclear weapons under International Inspection, and use money saved from military budget cuts for world economic development.

Levi Eshkol became Prime Minister in 1964 and Abba Eban got f LBJ's support for Israel's attack on its neighbors. June 8, 1967, and LBJ covered up Israel's attack on the USS Liberty on June 11, 1967 killing 32 Americans, wounding 174/

The murder of Abraham Lincoln in 1865 was done by John Wilkes Booth. But it was Simon Wolf, head of B'nai B'rith and Confederate spy ring in Washington, who appealed to Booth's vanity by saying if Lincoln were dead, the war could begin again and *this time the South would win.* After Booth killed Lincoln, Wolf ran away to England where British Intelligence assigned him to Cairo. In 1875 he bought the Khedives shares in the Suez Canal with Rothschild money which Rothschild sold to Queen Victoria *at a discount for a seat in the Parliament* and Wolf was appointed American Consul to Egypt in 1881 during which time Egyptian nationalist were accused of terrorism and Britain took Egypt.

Wolf used Booth to bring down Rothschild money from Quebec, Canada, that Wolf told him *was to rebuild the South.* But in reality B'nai B'rith *Carpetbaggers* came into the South to buy fertile land in the South for pennies and acres for Nathan *Rothschild.* Those who refused to sell were killed by mobs that burned their homes, raped the women, and their land *sold for taxes* to the Carpet. Lt. General Nathan Bedford Forrest created the *invisible army* to stop the killings and that became the KKK which was then used by the B'nai B'rith to *re-enslave African Americans* as *tenant farmers* to work on the huge *corporate farm made for Nathen Rothschild.*

Rothschild gave Simon Wolf the money to buy up the Egyptian Khedive's IOUs he gave the casinos of Europe to cover his gambling debts. Wolf then forced the Khedive to turn over his shares in the Suez Canal and Rothschild who offered those shares to Queen Victoria at a *discount* for a seat in Parliament, Benjamin Disraeli, a Jew, was Prime Minister.

1996, Mayor Rudy Giuliani creates the OEM (Office of Emergency Management to assess anthrax and Sarin gas attack potential wish Jerome Hauer as director. In 1999, FAA fined Logan Airport and Port Authority of New York $178,000 for 136 security violations showing they were ripe to carry it off

The PNAC debuted in 1976 in Washington D.C. as a *non-profit think tank.* Its *Statement of Principles* says: "It is necessary for the United States to build on the successes of this past century to ensure *our* security and *our* greatness in the next century." PNAC's leaders say American leadership is good for America by:

1. Increasing defense spending significantly to carry out global *responsibilities* and *modernize* our armed forces.
2. Strengthen our ties to *democratic allies* and challenge regimes *hostile* to our *interests* and *values.*
3. Promote political and *economic freedom* abroad?
4. Take responsibility for America's *unique role* in preserving and extending *international order* friendly to our security, *our prosperity, and our principles.*

Regime change in Iraq was a *top priority* with PNAC. They said Saddam Hussein had Weapons of Mass Destruction which he had destroyed to avoid a war with the US.

Dual American-Israeli Richard Perle, wrote a report titled; *A Clean Break: A New Strategy for Securing the Realm* and personally delivered to Israeli's Prime Minister, Benjamin Netanyahu and is to serve as a guide for American policy.

20

In 1998, Perle, Paul Wolfowitz (future World Bank president), R. James Woolsey, Elliot Abrams, dual American-Israeli citizens holding high offices in American government, John Bolton, and Robert Zoellick (who replaces Wolfowitz as World Bank president) demand President Clinton to remove Saddam Hussein: *"We can no longer depend on our Gulf War partners to continue to uphold the sanctions to punish Saddam as he blocks or evades UN inspections. American policy cannot continue to be crippled by a misguided insistence on unanimity."*

It was PNAC that claimed Saddam was hiding Weapons of Mass Destruction from UN Inspector and that Iraq's proximity to the oil fields of the Persian Gulf makes Saddam a threat to U.S. Clinton imposes sanctions through the UN but PNAC says sanctions are ineffective and send letters to Newt Gingrich and Trent Lott, urging Congress to act and support the *Iraq Liberation Act of 1998 (H.R.4655),* which Bill Clinton signed into law. PNAC focuses on how to use American military power to *guarantee success of America's leadership of the world.* Newt demurred.

On November 16, 1998, PNAC pushed to remove Richard Butler as head of the UN inspections. PNAC member William Kristol wrote an editorial in his online magazine, *The Weekly Standard* calling for sustained bombing and missile attacks. Kristol and Paul Wolfowitz proposed establishing a *liberated zone* in southern Iraq to make a safe haven for opponents of Saddam to organize a *credible* opposition to Saddam that the U.S. military will protect from the air and ground. Kristol through the *Emergency Committee for Israel* donated *one million dollars* to Senator Tom Cotton asking him to sabotage any *nuclear deal Mr. Obama would make with Iran.*

Some 80,000 Iranian Communists were enticed into Iraq to fight against the Islamic Republic of Ayatollah Khomeini. The

CIA trained and gave artillery, tanks, and finance then gave them over to Saddam to use in fighting Iran.

The CIA overthrow of Iran's democratically elected Majlis and Iran's Prime Minister Dr. Mohammed Mossadegh in 1953 was carried out by CIA agent, Kermit Roosevelt, grandson of Teddy Roosevelt, created the fictitious story that Dr. Mossadegh was *soft on Communism*. Norman Schwarzkopf got the Shah to sign a contract to allow American, British BP, and Dutch Shell Oil Companies to exploit Iran's oil. Shell was included because it is owned by the Rothschild family. The total price for overthrowing Iran's democracy was one hundred thousand dollars. As the CIA had no way to send the $100,000, an Iranian Jew lent the money. Workers in Tehran's bazaar were promised a hundred dollars to make a Communist demonstration for Hollywood. As the workers would be *Hollywood Stars* for an hour, they were jubilant and unaware *they were to be attacked and killed by irate Iranians angry that Mossadegh had not protected them from Communism.* But the attackers were from Iran's organized crime ad they killed all of them which is to drive home the idea that this was not staged.

William Kristol and Robert Kagan of PNAC wrote that another **Pearl Harbor** was needed to mobilize America for a war against terrorism. Franklin Roosevelt told his advisor Don Smith that he embargoed oil to Japan and threatened to go to war with them if they went to Sumatra to get oil for their planes, trucks, tanks the Japanese to get oil. British Intelligence broke the Japanese code and told Roosevelt the day and dawn attack on Pearl Harbor. The Japanese Ambassador came Sunday morning, December 7, 1941, to deliver a declaration of war. Roosevelt waited until he got the phone call to say *Japan stabbed America in the back.* Roosevelt sacrificed 4,000 American lives and some old ships to get America in WW II to save Britain because it held the League of Nations Mandate to make Palestine the Homeland for the Jews..

Lee Harvey Oswald was set up as the *lone gunman* who *killed* President Kennedy. Oswald joined the US Marines and and was sent to the military language school in Monterrey, California to learn Russian. He was to be flown over Russia and parachute down to connect with what was left of the German spy ring in Russia run by SS General Rainer von Gehlen. But after Gary Powers U2 went down over Russia, the plan was cancelled and Oswald was sent to Russia as an American wanting to live under Communism. The Russians took him in but assigned him to a motorcycle factory in Minsk which isolated him so it was all for nothing.

## Mossad Colonel Mike Harrari: 9/11 Mastermind

The author and around fifty men were gathered in the dining room of a hotel on Dong Feng Xi Road in Guangzhou the night of 9/11 (morning in New York), while the Mastermind of 9/11 was in his home in Bangkok. At 9:40 pm Thailand time, black smoke was reported as coming out of a gaping hole in the North Tower created by *a plane* that hit the tower but was in reality from explosives in the building.

As the North Tower burns a video showing an image of a plane flying past it and then making a sharp turn to hit to the South Tower and a gigantic fireball engulfing the South Tower is different than the explosions in the North Tower. Veteran CIA pilot John Lear says no plane hit either tower. No titanium jet engine has been found at the WTC or Pentagon as it would need 3,280 C of heat to melt it.

The morning after 9/11 in Bangkok, Thailand, which was September 12, 2001 at 6 am in Bangkok, Thailand, on September 12, 2001 at 6 am in Bangkok, Thailand, which was evening time in New York City, September 11, 2001 and just an hour earlier, Larry Silverstein had ordered the New York Fire Chief to *pull (demolish)* Building 7. Larry had to destroy his own building because it contained the *backup files of the missing 2.3 trillion dollars Dov Zakheim stole.*

Mossad Colonel Mike Harrari called his friend, Dimitri Khalezov, to come for breakfast in his home in Bangkok, Thailand on September 12, 2001 and bring nothing but *top shelf liquor.* Dimitri Khalezov was a highly trained Soviet officer of the "military unit 46179" known as "the Special Control Service" of the 12th Chief Directorate of the Defense Ministry of the Soviet Union. is an expert on nuclear weapons whose job it was to check on nuclear weapons in all countries around the world. He was to report if nuclear weapons were being used anywhere in the world and he covered the muni-nuke bombing in Bali in 2002. The Indonesian government executed three *extremist Muslims* but as they had no access to mini-nukes used in the bombing it was most likely Israeli agents who flourish in Indonesia, Malaysia, and especially Thailand where they buy the drugs from the Golden Triangle and supply Thi girls for sex.

Mike revealed on the morning of September 12, 2001 to Dimitri and his son, Michael, that he was the Mastermind who planned 9/11 and said that September 11, 2001, *was the greatest day of his life.* 9/11. It took years to plan and millions of dollars to finance and cooperation with people in the FBI, CIA, and Pentagon with top security clearances to collaborate with Israel in killing 3,000 people, cover up the theft of 2.3 trillion dollars, and steal 850 million in gold from bombproof vaults of the WTC. It was not the first time Israel killed Americans but it was the greatest number.

Dimitri has also written that the US Embassy bombings in Kenya and Tanzania Africa were mini-nuke bombings as well as the 1993 bombing of the WTC North Tower which the FBI under Louis Freeh sent Sheik Omar Rahman and his followers to prison on evidence the FBI fabricated.

In 1984, Peter C, Goldmark, Executive Director of Port Authority of New York, created the Office of Special Planning (OSP) *to evaluate vulnerability to terrorist attacks on the WTC.* In 1985 it called the WTC *a most attractive terrorist target* speculating a *time bomb laden vehicle parked in the basement parking area of a twin tower.* SAIC (Science Applications International Corporation) endorses it.

Sheik Omar Rahman and his followers were accused of the February 26, 1993 bombing of the North Tower. Kroll's deputy chairman, Brian Michael Jenkins then produced a report of a plane *deliberately* hitting a tower as a *possible terrorist threat.* Then in 1995, Timothy McVeigh, worked for the CIA and given the assignment to inculpate the independent militias in various states that Israel views as anti-Israeli with a van loaded with nitrogen fertilizer that could not have shredded reinforced concrete.of the Murray Building in Oklahoma City. The 911 attack on the twin towers followed Jenkins' analysis.

Dimitri added that a mini-nuke was used on the Murray Building. In that bombing, the steel reinforced concrete was shattered leaving the steel reinforcing rods dangling like spaghetti. In WW II, the US used old flying fortress bombers filled with 14 tons of explosive and flew them into steel reinforced concrete bunkers covering German Submarine pens and airports and these could not even make a dent or a crack in them.

The 2.3 trillion dollars Dov Zakheim stole was greater than the 1.9 trillion total IRS revenues in 2001. In addition to the 2.3 trillion, there was 850 million in gold taken from bombproof vaults of the WTC. It was not the first time Israel has killed Americans but was the greatest number but Israel has always been forgiven. Harrari's boast or admission was in a ***Veterans Today*** article entitled, *Staff Writer Dimitri Khalezov Talks on 9/11.*

Los Alamos National Laboratory (LANL), is where nano-thermite was developed. Danish scientist, Niels Harrit, has stated that he has found active thermite material in dust samples collected from the collapsed WTC towers.

Mike Harrari was the Mossad chief of Central America operating from Panama and ran its operations from Panama until President George H.W. Bush invaded Panama to remove *pineapple face* General Noriega who told the CIA that it was Torrijos who supplied weapons to the Nicaraguan Sandinistas for their defense against the CIA supported Contras as the US pit and arms embargo on Nicaragua. Torrijos had been involved in drug trafficking with General Lopez Arrellano of Honduras who was toppled from power by a coup from inside the Honduran Army. That was acceptable but arms to the Sandinistas and the CIA put a bomb disguised as a radio and in Torrijo's helicopter and Mike Hararis friend, Noreiga took over.

The Mossad was against the Sandinistas bnnecause they nationalized land given free to Jews by dictator Somoza Debayle, who was put in power by the US Marines. They were to split the profits 50/50 with him 50/50 and not taxes. The Mossad was out to destroy the Sandinistas and replace them with someone who would listen to Israel. Mike Harrari backed the Contras and President Reagan put White House military attaché` Lt. Col Ollie North in charge to help the Contras. Ollie ambition was to make General, and he obeyed.

Lui Galindo (real name Jerzy Rubinsky trained by KGB in Kiev, Interpol in Paris, and Mossad in Israel. He wrote the, report accusing the Sandinistas of constructing airfields all over Nicaragua for an aerial invasion by Fidel Castro. It was a lie as the airfields were made by sawmill operators and owners of large haciendas so they could fly to their haciendas or sawmills and avoid the country roads that were in bad shape.

The Sandinistas did not make those air strips. That report was used to put Mr. Reagan against the democratically elected Sandinista government. Israel gave weapons to the Contras and also gave Kurds in Iraq weapons to attack Saddam Hussein. But

26

when the weapons crossed into Turkey, Turks complained and the Mossad pulled out of Iraq leaving the Kurds to suffer and Saddam used poison gas on the Kurds to avoid losing troops..

.      In 1978, Iraq's leader, al Bakr, made a deal with Hafez Al-Assad to unite Iraq and Syria into one country to form a central nucleus to unite all Arabs.    Al Bakr would have the higher position and Hafez Assad the lower.  Israel forced al Bakr to step down and had the CIA put Saddam in power as Israel considered him a *moderate*.  Israel then gave Saddam with a list of 200 Baath Party members *and told him they planned to kill him.*   So of course Saddam killed them.

Harrari was known in Panama as *Mr. Sixty Percent* because that's how much he asked for any business deal he helped between Israel and Panama.    He and Noriega split the 60%. The US invasion of Panama began at 0100 on December 20, 1980. Mike Harrari and Israeli colleagues were caught at the airport by the 82nd Airborne before they could get on a plane to take them out of the country.   But after a call to an American politician, the 82nd Airborne helped Mike get on the plane that flew him back to Tel Aviv, Israel.

The shredding of steel reinforced concrete and turning it into dust was like the Oklahoma City bombing done by Timothy McVeigh.   McVeigh and nine other men undergoing *Special Forces Training* at Ft. Bragg, North Carolina were hired by the CIA to do drug trafficking to finance covert operations and government sanctioned assassinations.   The nitrate bomb in his van would not have shredded the reinforced concrete of the Murray Building to shred it like spaghetti.  The Murray Building was wired for demolition and mini-nukes were used which shredded the steel reinforced concrete as planes loaded with 15 tons of explosives and flown into German reinforced concrete bunkers couldn't make a dent.  A van filled with nitrogen fertilizer parked across the street from the Murray Building was to be blamed on Muslims but it was Timothy McVeigh, a Bronze Star hero of the Gulf War who was hired by CIA and assigned to do the bombing he killed Americans.

*Terrorism experts* Steve Emerson and Dr. Ron Hatchett *came immediately on television to announce that this was obviously the work of Muslims.* President Clinton *also came on television to tell the nation that Muslims did the Oklahoma City bombing* and then apologize later to say they weren't.

Dimitri Khalezov was involved in the implementation of the joint US-USSR treaty called the *Peaceful Use of Nuclear Explosions.* That allowed the use of nukes for the demolition of tall buildings like those of the WTC. The New York Port Authority insisted the architects of the WTC provide a means for demolition of the buildings when their useful would inevitably end. The firm of Controlled Demolition Inc. (the Loizeaux family) and it came up with the plan to use mini-nukes and thermite, which was how 9/11 was done. Mini-nukes in the basement and thermite on the corners of the twin towers to dissociate ties to the center. Controlled Demolition was given the job to remove the 9/11 debris but who filled the craters created by mini-nukes?

The steel used in the construction of the WTC was *tested* at Eglin Air Force Base in Florida. Eglin reported that the steel was inferior and not fire-resistant steel required in all high rise construction. Eglin is where the CIA had a micro-chip inserted Timothy McVeigh's buttocks so they always knew where he was. McVeigh did the Oklahoma City bombing to blame on Muslims so Congress would not lift the arms embargo on Bosnia to defend itself.

The seven-ton steel Granit cruise missile used to hit the Pentagon came from the wreck of the Russian submarine, Kursk. The missile weighed seven tons, fueled with oxygen peroxide and designed to fly just above the surface of the ocean and sink an aircraft carrier at the water line. It had a 500 kiloton nuclear warhead - 50 times more powerful than the bomb dropped on Hiroshima. It Granit missile flew at 2.5 Mach or 3,280 kmph at sea level. NORAD above Colorado Springs spotted the missile at 9:37 (six minutes before it hit the Pentagon and sent out an alert. V.P. Dick Cheney was running a practice drill of an air attack on the United States and all the planes were under his control. But from a drill to what was reported as *reality*, no plane was sent to intercept. *Wikipedia's* article on 9/11 has Hani Hanjour as the pilot crashing the airliner with 63 on board and 125 in the Pentagon dying.

It was a blot on the *Wikipedia* reputation to accept the disinformation coming from CIA and NIST reports. Marine Corps heavy helicopters from Quantico flew in low to the Pentagon to knock down light pole to simulate *United Airline flight 77* and Muller's FBI doctored the security tapes that showed a missile hit that Pentagon office killing 29 accountants and destroying the records of how Zakheim stole the money.

The missile went on to destroy Naval Intelligence in ring three. As the detonator on the missile did not work the whole section, *was brought down by explosives to bury the missile.* All plane parts as evidence have disappeared.

Dimitri Khalezov was head of Soviet intelligence on nuclear weapons around the world before the Soviet Union broke up December 25, 1990. Dimitri was a friend of Colonel Mike and in Bangkok to help fellow spy, Viktor Bout, fight an extradition warrant to the US over a sale of weapons to FARC of Colombia. Dimitri says the charge is a cover up to something more sinister.

Why was Harrari living in Bangkok? Thailand is where the CIA set up the Golden Triangle to finance a Chinese Nationalist General. Israeli's Mossad and American Jews marketed the opium from the Golden Triang since 1949 and Jews have a free hand to live or do business in Thailand after they showed the king how to make more money from his money and tourism, rugs, prostitution, and more flourished in Thailand for rich kids like

Van de Sloot who had a hangout after he left Aruba where American teenager in his company disappeared.

## Effect of September 11, 2001 in China

On September 11, 2001, the author was on the other side of the world from New York City attending  wedding of a *Palestinian couple,* while one man posing as a Muslim was video-taping the event as *Muslims celebrating the news of 9/11.* The truth is that this grotesque mass murder was to cover up 2.3 trillion dollars stolen by Dov Zakheim and multi-million dollar gold heist from the basement bomb proof vaults of the WTC.

On May 4, 2001, Mr. Bush appointed Rabbi Dov Zakheim comptroller of the Pentagon May 4. 2001 and given control of three trillion dollars.  In just four months. 2.3 trillion dollars disappeared and the world's worst currency, the Israeli shekel, began to rise and in two years, the shekel became a currency matching the Canadian Dollar, Australian Dollar, Euro, Yen, Reminbi, and US dollar.

Since Israel's independence in 1948, the shekel was devalued daily.  Israel's balance of trade and finances was a shambles.  President George H.W. Bush guaranteed Israeli's ten billion dollar loan to finance illegal Jewish settlements on Palestinian lands to get the Jewish vote and financial help.  When Israel defaulted, the 10 billion was paid by the US and was separate from the billions in military aid, oil subsidizing, and economic aid.

President George W. Bush said it was America's freedoms that allowed Arabs to infiltrate America to carry out 9/11 and called for laws to take away America's freedoms and rights guaranteed under the Constitution.  When the Senate Majority leader, Tom Daschle, and Senator Leahy opposed him.

So President Bush said, "The Constitution is nothing but a God damn piece of paper."

On September 18 just one week after 9/11, Bob Steven of the news media was killed by opening an anthrax letter. Then Tom Daschle's secretary was poisoned after opening an anthrax laden letter addressed to the Senator. Then Senator Leahy's secretary was next. Anthrax poisoning was to be blamed on Muslims as well as the WTC suicide bombings.

On October 26, 2001, George W. Bush signed the Patriot Act. *Anyone accused of associating with anyone known as a terrorist can be accused of being a terrorist, subject to arrest, seizure of property, bank accounts, and imprisoned without representation by an attorney and detained indefinitely.* Susan Lindauer, a peace activist, was working on behalf of the United States with Saddam's Iraqis to destroy WMDs. A few days after George W. Bush took office in January 17, 2001, she was fired from her job. When she revealed that Saddam was doing all she had asked them to do, the Bush White House said they had evidence Saddam was hiding WMDs. When Susan denied that charge, she was arrested and imprisoned under the Patriot Act for consorting with terrorists. The court said she was insane and incapable of being tried. So the government offered to give her drugs so she could be tried. The court rejected this and freed her.

A Muslim professor at a Boston university wrote an open letter, *Wake up Call for America to blame* 9/11 on pro-Israel policies of America. He jumped to this conclusion before the full facts were known. Then the Boston Marathon massacre was blamed on two Muslim brothers from Kirgizstan. Hollywood producer/director Nathan Folk has denounced the Boston Massacre as a Black Ops. Interviewed by Russian TV, Folk's story that a make=up artist was outing blood make-up on the *wounded* was very believable. Russian TV showed then videotapes two white surveillance vans parked on the street across from Folk's house and after this broadcast by Russian TV, the cable carrier cut Russian TV from its service.

12 years after 9/11, the ISIS Islamic State appears in northern Iraq and Syria conducting psychological warfare. A British Intelligence Report revealed a Mossad-MI6 plan that was first called *Operation Britannique*, then *The Hornet's Nest,* and was to attract extremists from all over the world to create what is now ISIS. They were to use the psychological warfare the Pentagon taught to Latin American military dictators back in 1979 to 1986 that left headless corpses in Chile's in river beds in the dry season and in El Salvador's mountain villages. .

Britain is home to two million Muslims from all over the British Commonwealth. Britain's MI5 approached four Muslim men of Pakistani ethnicity to participate in an *exercise* to test British security measures. They volunteered as they wanted to show their loyalty and died July 7, 2005 in London's *Muslim orchestrated attack on London's bus transportation system as suicide bombers.* The families of the men have denied they were suicide bombers or fanatical in any way.

After the bombing of London buses, the British police were given guns and a Brazilian living in London who looked *Middle Eastern* was shot in London's subway system. When it was revealed he was Brazilian and not a Muslim from the Middle East, the police claimed he jumped the Underground turnstile and menaced the police in a manner that they said they suspected he was carrying a bomb.

December 22, 2001. British convert to Islam, Richard Reid, who tried to ignite explosives in his shoes by clicking them together. When Reid tried to light matches to make the shoes explode he was subdued and American Airlines flight was escorted by two Air Force fighter planes to *Logan Airport*. Ten ounces of plastic explosive were found in the soles of his shoes. On May 5, 2015, *everyone* had to take off their shoes at the security checkpoint at O'Hare International in Chicago.

On Christmas Day, 2009, Umar Farouk Abdulgssem Maballah a 23 year-old Nigerian student boarded a Northwest Airline flight from Paris to Detroit. He was subdued and flight diverted to Logan International where CSI, an Israeli security company, provided security at Schiphol, Logan and Dulles International. The Mossad has an office at El Al in Schiphol from which it runs its drug operations in Europe/.

On March 1, 2014, *The Telegraph* published the news that former Iranian intelligence officer, Abdolgassem Mesbahi, defected in Germany and told that Supreme Leader Ayatollah Khomeini ordered a bombing of an American plane "to do what the USS Vincennes did to Iranair flight 655 to Bahrain that killed 290 innocent men, women, and children. The Vincennes used a high tech missile on Iranair and the bombing of Pan Am Flight 103 was not carried out by Hamas, Libya, or the PLO, but by *The Populon ar Front for the Li beration of Palestine.*

On July 2, 1988, the Liberian tanker, Stoval, was surrounded by 13 Iranian gunboats and the US Navy frigate, Montgomery, sent a call to the USS Vincennes, a hi-tech cruisers at the Straits of Hormuz where Iran had based lightly armed speedboats. The US placed an embargo on Iranian oil and Iran threatened to sink oil tankers from Kuwait that was donating money to Saddam Hussein's war with Iran. The speed boats were no match for the Vincennes's radar guided missiles

The next day, Captain Will Rogers III received notice from his radar technician that a plane was flying from Iran toward them and Rogers ordered it to be shot down on the supposition it was an Iranian Air Force plane. So 290 innocent people died. Abdulbeset al Mahmoud el Megrahbi, Libya's intelligence chief was blamed to incriminate Libyan leader Maumar Gadaffi, who gave him sanctuary but he surrendered to the British courts and was sentenced to prison in 2001. Maumar Gadaffi paid the families of the victims $2.7 billion dollars. In 2009, el Magrahbi was diagnosed with terminal prostate cancer and allowed to return to Libya where he was given a warm homecoming. The news media made a lot out of that but many of the families of Lockerby said that thought he was innocent..

The US refused to apologize. In San Diego, California, the SUV of the wife of the Captain Will Rogers III of the USS Vincennes. Captain Rogers has maintained he was acting correctly in his decision to shoot down the Iranair plane.

A group of British ethnic Pakistani-Indian Muslim teenagers thought it would be clever to see if toothpaste, skin creams, lotions, liquids in plastic bottles, etc. could be used to make a bomb while in flight that could then be used to blow up a plane. Fanciful stuff and one man wrote a story on the news media of how his wife mixed some household cleaning fluids to make poison gas or a bomb to prove it was possible. All hype and that was all it was.

Hundreds of millions of passengers have had perfumes, saving lotion bottled water, medicines, toothpaste, and nail clippers taken from them in the name of airline security?

What is causing Muslim immigrants to act this way? When they came to Britain or other western country their goal was to improve their lives. Muslims like Britain and the US. If they didn't they would go back home. It *was* ridiculous as these teenagers were acting silly, caught as potential terrorists by the British police and now it is a security check.

The author remembers in high school that he and a classmate thought it would be fun to make nitrous oxide that could be introduced into the ventilating system of the high school and have everybody laughing. We talked about how to do it but never did it. Nitrous oxide won't hurt anyone and dentists used it as a pain killer. The bad thing about laughing gas is that you get a headache afterward.

Back to the subject of Middle Eastern business men in China being set up to be *involved in 9/11,* the author was teaching English at a private school in Guangzhou when a friend from

34

Yemen introduced him to a dress manufacturer who wanted him to teach his cousin English. The author was intrigued but the bad side of working for him was the job did not come with an apartment.

The author was 71 at the time and the boss was hard-nosed about working six days a week for ALL employees. It soon became apparent there was another reason he was hired. His Chinese girlfriend worked with him for seven years. Hiring the author as a *manager* was to keep his girlfriend submissive while he carried on with girls looking for a job. His girlfriend was always nasty with the other girls to show that she was the boss's equal and not just his bed warmer.

The boss never went for Friday prayers at the Mosque in Guangzhou which with his proclivity to sleep with all girls he could had the author wondering if he were truly Muslim. He told the author stories about the secret society of the Roshaniya/Illuminati and their sexual practices in Syria. The author listened but he had yet to prove he was a reliable source.

The office staff went to the 2001 fashion fair at the Hong Kong convention center and the boss got millions of dollars in orders. He was ecstatic. Back in Guangzhou, he introduced the author to Nezzar, a *Palestinian,* who invited the author to a celebration of his brother's wedding. The author didn't know Nezzar or his brother but accepted but as he wasn't familiar with the hotel on Dong Feng Xi Lu Road where the celebration was being held he got lost. When he finally arrived it was 8:30 pm and everyone was waiting for him. Nezzar began beating on a drum and led the way to this big room where there was to be dancing to celebrate his brother's wedding. Already seated at a table was the author's boss, his girlfriend, his cousin and his cousin's Chinese wife. The boss was fiddling with an expensive camera the author had never seen him with before and when asked about the camera he was cryptic and didn't give a direct response.

Some Arab music started playing and some of the men got up to dance and the author joined them to show he was still young at 71. Nezzar seemed to be extremely happy about the author joining with the others. But as he danced his face took on a satirical look that shook the author. About ten O'clock, two men came running out on the stage jumping up and down and very happy and shouting in Arabic. The author looked at his boss for a translation but he just smiled and fiddled with his camera. No one in the room was jumping up and down or giving hi-fives.

After a while, the music resumed and everyone went up to congratulate the new couple who were now seated very sedately on a sofa on the stage and then everyone went home.

The next morning, the streets of Guangzhou were quiet which is unusual in China as there is more noise from people talking than the noise from the traffic. On the bus, people were quiet. It was only after the author got to the office he learned about 9/11. As he had taken photos, he gave his film to the bookkeeper of the company to be developed. He didn't see the boss taking any pictures with his expensive camera but as men from the celebration began coming to his office and as the author doesn't speak Arabic but understood they were quite concerned. There was something that shook the author when the boss implied that as I was American I might be CIA. Later in the week, the boss sad he was having trouble keeping his license to manufacture his evening gowns and wouldn't be able to get the work permit promised and told the author to go to Hong Kong to renew his visa every three months.

9/11 impacted the economies of Middle Eastern and European countries and that hit my boss in China. Clients who placed millions of dollars in orders at the Hong Kong fashion fair were cancelling. The author's job was reduced to teaching the staff American business English, then the boss informed the author he had to let me go.

By now, the Chinese intelligence was watching all Muslims in Guangzhou and it was so tight that hundreds of factories in Guangdong were shutting down. 9/11 was definitely impacting the Chinese economy. But now the author was working with a private school and had to go to Hong Kong every thirty days for a new visa.

After picking up his passport at the Hong Kong visa office the author went Ho Hum station in Kowloon to get the train to Guangzhou. But the last train had already left. So he rushed to the bus station only to be told the last bus to Guangzhou had left. But as there were many people waiting to go to Guangzhou they brought two more buses.

After taking his seat the author noticed a man looking sharply at him and asked the man if he spoke English. He said he was a professor and proficient in English. A girl behind him was looking at the author as if he were some cockroach to be stepped on. Then the Chinese woman sitting behind the author spoke up and said she was from Australia and informed him that Muslims in Australia were dancing in the streets the night of 9/11 and celebrating. The author replied that it wasn't right but learned later that Australian archive TV videos of Australian Muslims celebrating the end of Ramadan were shown on TV *as Muslims celebrating 9/11*.

The author began to piece together the sharp eyed man and the unpleasant girl sitting behind him as well as the reaction of the girl issuing my visa at the visa office in Hong Kong. When the author handed her his business card as an editor of books, she asked sharply, "What kind of books do you edit?" The author replied he had edited the books of a well-known author and added in a loud voice, "He's friend of China, for Pete sakes!" The girl made a sour face and stamped his passport. The author then learned from a former co-workers the boss's expensive camera was video camera and he took videos of the author and the others dancing and turned them over to the Chinese police to show

Muslims as dancing the night of 9/11 which was the morning of 9/11.

Before the author left the fashion business, his old boss told him that a client from Astrakhan, Russia, informed him that a Russian newspaper had published that Russian Intelligence learned that the Black boxes recovered from *the planes* involved in 9/11 had nothing on them. That was a decoy was a decoy story as there were no planes. Six months after 9/11, Russian and Chinese intelligence discarded the story that 9/11 was done by Al Qaeda who are not terrorists but a resistance group trained by the CIA to fight Russians in Afghanistan. Osama bin Laden paid their way to go to Bosnia to fight Serb terrorism, the CIA labeled them terrorists, called them *Al Qaeda* to legitimize the invasion of Bosnia by Serbia.

After the author left China, a Chinese girl accused his old boss of rape and he spent two years in a Chinese prison. He lost a lot of weight pounds and doesn't look the same.

9/11 needs an Autonomous International Inquiry to examine all evidence seized by the FBI around the Pentagon and elsewhere that have been kept hidden from the public. It is obvious that Israel would do all it can to block an investigation and why Mr. Obama has not allowed.

President Bush used Eglin Air Base to test the steel used in the construction of the WTC. The Eglin report said the steel used in the construction was of inferior quality but United Laboratories of South Bend, Indiana (noted for its integrity) said the opposite.

As all steel used in hi-rise construction must meet high standards to avert a collapse of a hi-rise building in case of a fire. Larry Silverstein built the WTC complex 1970-72 and if he used inferior grade steel in its construction of the WTC then the Fire Marshal of New York City should have convened a grand jury that would have to indict Larry Silverstein on 2,780 felony charges of manslaughter (3rd degree murder). So if Larry

Silverstein used inferior steel to build the WTC complex in 1972 he should been indicted for manslaughter by the New York by the New York City Fire Marshal

## Ex-CIA Pilot Says No Planes Hit WTC Towers

John Lear holds more certificates for flying all kinds of aircraft than any other pilot licensed in the USA, He flew 34 years for the CIA and has submitted an affidavit that will the government must refute his affidavit or his affidavit will become the record. John is the son of Bill Lear who makes business jets and has testified it was physically impossible inexperienced pilots who trained on one engine propeller driven to take over the controls of a Boeing 767, like Flights AA11 and UA175 fly them into the Twin Towers on 9/11. John says no plane hit either of the towers and the white smoke from the South Tower came from explosives.

John Lear stated unequivocally that "No Boeing 767 airliners hit ether of the Twin Towers as alleged by the US government, the news media, and NIST. A real Boeing 767 would have begun *telescoping* when the nose hit any of the 14 inch steel columns which are 39 inches apart in the center of the Towers that encased 169 elevators. 'The tail section would have stuck out of the tower and fallen to the ground."

The engines on impacting the steel columns would have maintained their general shape and either fallen to the ground or been recovered in the debris. 'No Boeing 767 could attain a speed of 540 mph at 1000 feet above sea level as the *'parasitic drag* doubles with velocity' and *'parasite power'* cubes with velocity. The fan portion of the engine is not designed to accept the volume of dense air at that altitude and speed and 'The debris of the collapse should have contained massive sections of a Boeing 767, including 3 engine cores weighing approximately 9000 pounds apiece which could not have been hidden. There is no piece of any of these massive structural components from either 767 at the WTC. The complete disappearance of the 767s is impossible.

There is the picture of a girl waving from the hole *Flight 11 made* (see p.43 the photo) and the four floors of the North Tower where flight 77 hit were in office space rented to companies that made billions in profits from Pentagon contracts.

John Lear's affidavit was dated 28th January 2014 and part of a law suit being pursued by Morgan Reynolds in the United States District Court, Southern District, New York. John Lear previously filed a Request for Correction with the US National Institute of Science and Technology in March 2008 based on the lack of pieces from a Boeing commercial jets, and has produced evidence no plane hit the WTC towers.

The 9/11 Truth movement initially rejected his 'no-planes' theory, but after scientific analysis, they have his accepted his explanation. Unlike any other form of statement, an affidavit becomes truth in law, if it is not rebutted. It will now be up to critics of the theory to present their evidence and analysis point by point. If they do not, his sworn statements in the affidavit will prove that the US Government NIST report is wrong. But so far nothing has been done.

Lear is a 65 year old retired airline captain and former CIA pilot and has over 19,000 hours of flight time. He states that the inexperience of the accused "pilots" who trained on single engine propeller could not control the descent of a large multi-engine jet plane into New York City on a straight course. The difficulty in controlling heading, descent rate, and descent speed within the parameters of a *controlled flight* is beyond the capabilities of the alleged pilots certified to fly single engine propeller planes.

'It takes a highly skilled pilot to interpret the *EFIS* (Electronic Flight Instrument Display) display, which none of the alleged hijackers would have been familiar or received training on or known how to use the controls the ailerons, rudder, elevators, spoilers, maintain speed and manage to descend without crashing."

John Lear has flown over 100 different types of planes during his 40 years of flying, and holds more FAA pilot certificates than any other FAA certificated pilot. He flew secret missions for the CIA in Southeast Asia, Eastern Europe, the Middle East, and Africa between 1967 and 1983 as well as 17 years of working for passenger and cargo airlines as Captain, Check Airman and Instructor. Both United and American airlines are said to have never filed claims for losses with their insurance carriers and an air controller at Cleveland International has identified one of the American Airlines landing there in 2003.

WTC Security was handled by Securacom, a Kuwait-American firm and Marvin Bush, President George W. Bush's brother, was on the board of Securacom which also served Eglin AFB, located in Florida. Eglin is near another MacDill AFB, where Dov Zakheim sent *32 Boeing* 767 aircraft to Israel as part of the Boeing/Pentagon tanker lease agreement to replace Israel's aging KC-135 fleet with smaller, more efficient Boeing 767. After 9/11, Securacom changed its name to Stratosec and was delisted from the New York Stock Exchange in 2002.

The black smoke from came from oxygen starved fires and unlike the South Tower fire with kerosene burning on the outside of the South Tower. Then there is the photo of a girl waving from the hole in the North Tower before the black smoke came pouring out of that hole. How could this girl have survived if the plane's fuselage was completely consumed by fire as stated by the NIST report? It goes to support the evidence presented by Colin Alexander that the videos of a plane crashing into the South Tower were digitally altered to show a plane that did not exist. Colin Alexander's analysis of the plane's trajectory show that it varied with each vide. Its absolute straight flight pattern goes against turbulence that occurs at altitudes lower than 1,000 feet.

41

There is a video where President George W. Bush says he saw a video of a plane that hit the North Tower *before the second plane hit the South Tower.* Was that a Freudian slip he was in on the operation?

The story that *Flight 77* that hit the Pentagon was a Russian missile, Granit, designed to sink an aircraft carrier by flying just above the surface of the ocean and hitting an aircraft carrier at the water line fits the parameters of the missile that flew just above the surface of the ground and penetrated the outer wall and continued on to the Pentagon's 3rd ring to destroy the Office of Naval Intelligence. All physical evidence of plane parts at the Pentagon disappeared. Only photographs remain.

Flight 93 was a Boeing 757 that wandered off-course. *There was no take-over of the plane by hijacker* and was shot down in the panic that set in over the pilots failure to respond to radio calls. A cock-pit recording of a hijackers breaking into the cockpit with voices of the intruders and the co-pilot shouting, "Get out! Get out!" followed by a gurgling sound said to be blood going down his windpipe as his throat was bad *sound effects* as a slit throat does not penetrate to the esophagus. That the Pentagon *faked the tape* is obvious and shows collusion by submitting false evidence involving high levels of the Pentagon

A video clip of **Flight 175** turning to hit the South Tower showed a cylindrical shape on the bottom of the plane which is consistent with the size and shape of a cylinder carried on the bottom of a Boeing 767 refueling tanker. Three were given to Israel to refuel Israeli bombers and fighter plane on missions beyond the range of Israel's bombers such as targets in Iran, Turkey, Saudi Arabia, Russia, and the United States.

This information is to mislead many that flight 175, as pictured on the news media and official reports, was a refitted Boeing 767 tanker that was given to Israel but replaced by newer refueling tanks by Dov Zakheim. Showing this old tanker on the video with a Flight Termination System still attached was good psychological operation would convince many to say that the Boeing 767 was converted tanker to support the NIST report that it was flown by inexperienced pilots that only learned to fly a Cessna.

Both New York and Washington investigators credited 9/11 to inexperienced flight school students. "Since a refitted 767s tanker given to Israel by America and repainted to look like a commercial airliner, people boarding the plane would not be looking at the strange tank under the plane's body. I would have numbers painted on it and could carry both passengers and a fuel load. The flight tower would have the plane listed as Flight 175 and it would not be recognized as a refitted 767 tanker painted as a conventional commercial passenger plane.

But former CIA pilot John Marshal said that Evergreen rents planes to the CIA and paints them any color or Logo on the plane that the CIA. He and his two children and family dog were shot in what the California police called a double murder and suicide.

All flights involved in the events traveled very near many military installations, and appear to have been flown in a manner suggesting guidance and possible transfer of the control of the planes among the bases or a mother-ship nearby.

For three days nothing the heat from the collapsed twin towers nothing could be done to remove the debris as it was so hot. Even after two weeks the debris removal was bringing up molten steel glowing red hot.

Eglin Air Force Base in Florida chosen to test the quality of steel used in the construction of the twin towers. It is thoroughly corrupt. An example is when drug trafficker, Juan Ramon Matta Balesteros, was sentenced to two years in prison in 1970 on a passport violation, it was after he was caught by the DEA with 22 kilos of cocaine at Dulles International.

After a year, Juan Ramon *escaped* from Eglin Federal Prison and made it back to Honduras withou*t knowing English or*

*Geography of Florida.* He met Carlos Lehder of the Medellin Cartel at the villa of Roberto Vesco who stole 250 million dollars from Bernie Cornfield's *Fund of Funds (The Dreyfuss Fund)* that was bankrolled by 50 million dollars of Meyer Lansky's money to launder the Mossad's dirty drug money. Lehder marketed the Medillin Cartel's cocaine and the CIA and Mossad bought cocaine from the Medillin Cartel which was how the Cartel became big as it was protected by both the CIA and Mossad.

Carlos took Matta to Colombia and taught him how to fly planes. Ollie North and General Secord helped Juan Ramon buy army surplus *flying boxcars* that became SEATCO and flew cocaine into Mena, Arkansas where Bill Clinton grew up. Carlos was half German and half Colombian and had a four inch bronze bust of Hitler on his desk, but when Mike Harrari saw the bust of Hitler he blew up and demanded Lehrer be taken out and that ended Cartel. The DEA estimated Matta supplied one third of all the cocaine smuggled for Lt. Col. Oliver North.

Eglin Air Base was where the CIA had a microchip inserted in Timothy McVeigh's butt. McVeigh along with nine other men were undergoing special forces training at Fort Brag, North Carolina and were hired by the CIA to do government *sanctioned assassinations and drug trafficking to finance covert operations.* McVeigh was assigned by the CIA to do the Oklahoma City bombing was to be blamed on Muslims and failed only because Oklahoma State troopers picked up McVeigh for driving a car without license plates and his photo matched the composite drawing of the Oklahoma City bomber.

Dov Zakheim is a member of PNAC (Project of a New American Century) which advocates a *new Pearl Harbor* to push Americans to accept a more highly militarized American foreign policy and military build-up. PNAC does not comment on the disappearance of 2.3 trillion dollars that member Dov Zakheim *stole.* Should anyone with dual American-Israeli citizenship be allowed to handle top security or be elected to Congress? 29

44

Congressmen are dual American-Israeli Americans. Dov's grandfather was a top Communist in Russia while former Secretary of State, Madeleine Albright's father was a top Communist working in Communist Czechoslovakia.

When word of the first plane that hit the twin towers came to Logan International Airport in Boston, all Logan airport employees passed word around to make a record of what they saw and heard at the airport at the time the passengers checked in, boarded the planes, planes taxied to the runway, and time they took off. When the Israeli-American chief of airport security learned about the staff documenting what they had observed at the time the planes took off he told them to destroy the documents they had made.

That left him open to be charged with impeding an investigation of a felony. But he wasn't charged so it casts suspicion on the Security and management of Logan Airport. The evidence that no planes hit either tower is very convincing. The video showing the plane that hit the South Tower was a tanker was a decoy video to mislead and have a court reject all evidence.

## Who Was Dov Zakheim ?

Dov Zakheim's grandfather was born in the Ukraine in 1870. His name was Julius Zakheim (Zhabinka) who was participated in the 1905 Communist uprisings. The Bolsheviks were behind the Zionist Movement which was based on the creation of a Homeland for the Jews in Palestine.

Dov's father, Rabbi Jacob I. Zakheim was born in 1910 near Bilaystok, Poland. His family included Menachem Begin and Moshe Arens who later headed the Mossad. Menaghin Begin was a terrorist in pre-Israel and killed hundreds of British soldiers as well as Count Folke Bernadete, Cousin to the King of Sweden and International Red Cross inspector of all German prisons and concentration camps during WW II. The Mossad is connected with crime organizations around the world and is noted for assassinations all around the world that are made to look natural like the poisoning of Abdel Gamal Nasser in 1969.

President Nixon wanted to give Egypt nuclear power as the Aswan dam was not producing enough power for Egypt and unemployment might lead to war with Israel. He also had the UN allow Iran to develop nuclear power. He told Dr. Kissinger to tell Israel to make peace with Palestinians or further aid would be cut off. But Kissinger did nothing.

Dov's father immigrated to the United States and Dov was born in Brooklyn in 19 43. He attended exclusive Jewish schools, spent summers in Israel Zionist camps. He graduated from Columbia University in 1970 and then attended the University of Oxford in 1972 in England where he attended the London School of Jewish studies, 1973 to 1975 with courses on Jewish Supremacy, Advanced Bible, Talmud, Mysticism, the Holocaust, Anglo-Judaica, and Zionism and that led to his being ordained a Rabbi.

Dov was also an adjunct professor at the National War College, Yeshiva University, Columbia University and Trinity College, Hartford, Connecticut from 1975 to 1980. On joining the Reagan administration, He talked them into funding the development of the Lavi Fighter at a cost of $3 billion. It was a dud and Israel dropped it. The US cancelled $450 million in contract fees so Israel created a story that China was eager to buy the Lavi and Dov convinced President Reagan that to stop the Chinese from acquiring the Lavi, Reagan to give Israel $500 million for its lost $450 million in contract fees. Dov Zakheim had a wing F-16's and F-15 *classified as surplus* and sold to Israel for a fraction of their true value. At that time, 15% of all US Air Force jets were grounded for lack of parts, and left America less protected.

Dov is an ordained rabbi. For 25 years he influenced the defense policy of Presidents Reagan, Bush Sr., Clinton, and Bush Jr. Paul Wolfowitz, Elliot Abrams, Ben Wattenberg, and Richard Perle, Dov never served in the US military and all are *dual Israeli-American* and PNAC members hold high positions on American defense issues that failed on 9/11,.

From 1987-2001, Dov was CEO of SPC (System Planning Corporation) a high tech *analytical* consultant firm to the Office of the Secretary of Defense in Clinton's Administration. In 1997, Dov was on a task force for Defense Reform. In 2000, he was foreign advisor to President George W. Bush and the *Vulcans*, a foreign advisory team led by Condoleezza Rice that included Richard Armitage. Robert Blackwell, Richard Perle, Paul Wolfowitz (later expelled from the World Bank as president), Robert Zoellick (who replaced Wolfowitz in the World Bank), and Scooter Libby, were to brief Mr. Bush on military and foreign policy.

In September 2000, Dov became a member of the Council on Foreign Relations, the United States Naval Institute, on the editorial board of *The National Interest,* a three-time recipient of the Department of Defense's highest civilian award, Distinguished Public Service Medal, Adjunct Scholar of *Heritage Foundation,* Senior Associate at the Center for Strategic Studies, and published over 200 articles and monographs on defense issues. On May 4, 2001 he was appointed Under Secretary of Defense Comptroller and given 3 trillion dollars. Four months later on September 10, 2001, Secretary of Defense Donald Rumsfeld said 2.3 trillion dollars had disappeared and the next day, a Russian missile, hit the office where 56 accountants were going over the books to find out who got the 2.3 trillion dollars. The records were destroyed as were the back-up files in Building 7 when it was *pulled.*

On May 6, Booz, Allen Hamilton gave dov a job on the *Defense Advanced Research Projects Agenc.* Booze, Allen & Hamilton was owned by the Carlyle Group which hs ties to former president, George H.W. Bush and the Bin Laden family. The morning of 9/11, George H.W. Bush was having breakfast with the Bin Laden family to discuss Carlyle Group projects..

In 2008, President George W. Bush appointed Dov Zakheim to the Commission on Wartime Contracting in Iraq and Afghanistan. In 2010, Zakheim retired as a *Senior Vice President* of Booz Allen Hamilton. In 2011, Dov was Middle East advisor for Mitt Romney's 2011 presidential run. Mitt, Dov, and Larry Silverstein, are Benjamin Netanyahu's close friends.

## An Open Letter to President Obama

In an open letter published in the UK *Guardian,* 50 well-known individuals and watchdog organizations wrote that the Obama administration's record on secrecy and surveillance was a *disgrace, and there were more requests for information* under the *Freedom of Information Act* in President Obama's first term that previous years and the response was dismal. The Obama administration squashed more *legal inquiries into secret illegalities than any other president* and *a record of persecuting, prosecuting, and imprisoning those who came forth with information of corruption than any other president."*

*Five organizations* awarded Mr. Obama in a closed, meeting at the White House, ***"If the ceremony had been open to the press, it is likely that reporters would have questioned the organizations' proffered justification for the award, in contrast to the current reality."***

The Obama White House has refused to make its visitor logs public. The ***Obama's Department of Justice twisted the 1917 Espionage Act to press criminal charges in five alleged instances of national security leaks*** and his Administration initiated "more prosecutions under this act ***than all previous administrations*** combined." and "set a powerful and chilling example for potential whistleblowers through the abuse and torture of **Bradley Manning,"** (he was kept under *Suicide watch* for two months with his cell was lighted around the clock and prodded to wake up and respond to a guard every 5 minutes..

The Patriot Act took away the Constitutional rights and freedoms so President Bush issued an *executive order* to kidnap, kill, torture, and secretly imprison people in other countries, without charge. In detention centers, Mossad agents gave advice on tortur. They recruited and recruited *El Bagdadi.*

## Pictures of 9/11 and Gaza's 14 Years of 9/11

Larry's building 7 had the up files on the 2.3 trillion Dov stole.

Each tower had 90,000 tons back-of steel which is stronger than reinforced or asbestos concrete.

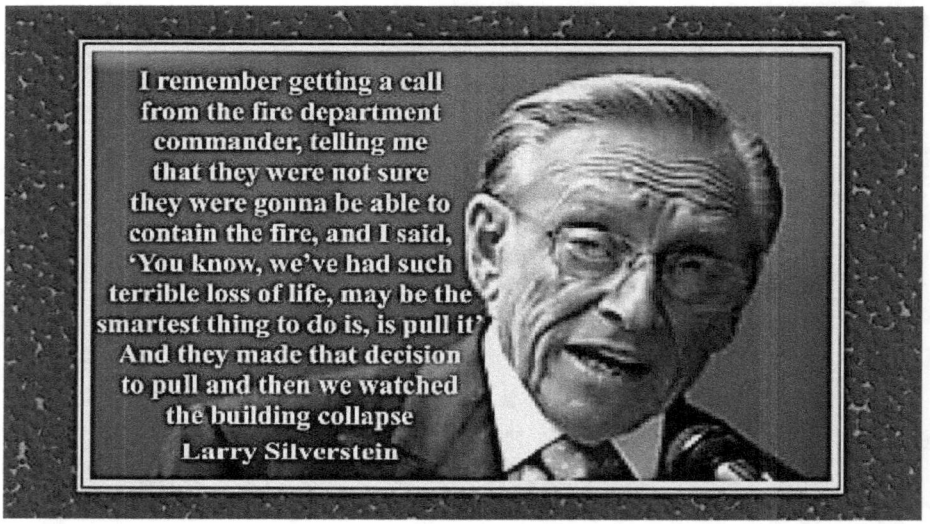

I remember getting a call from the fire department commander, telling me that they were not sure they were gonna be able to contain the fire, and I said, 'You know, we've had such terrible loss of life, may be the smartest thing to do is, is pull it' And they made that decision to pull and then we watched the building collapse
**Larry Silverstein**

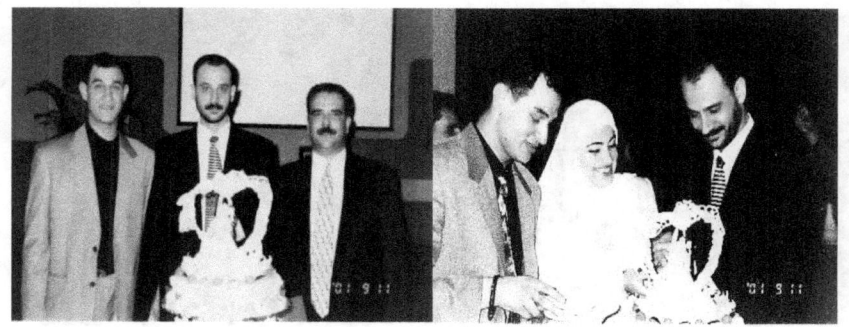

Fake wedding celebration in Guangzhou China, night of 9/11

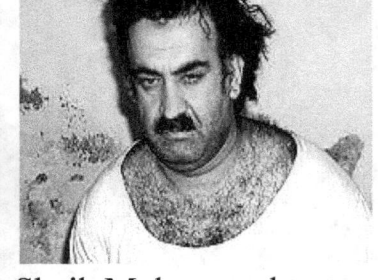

Mike Harrari died Sept 21, 2014   Sheik Mohammed tortured
admittted Mastermind of 9/11        187 times was innocent.

John Lear says no planes hit   Thermite and mini nukes used
In the perfect demolition of the twin towers.

67

Edna Cintron waving in North Tower hole 22 seconds before tower collapsed In a free fall as others jumped tom the roof

Dimitri Khalezov and Veteran's Today journalist.

An average of 22 veterans with PTSD commit suicide daily

These upper tory beams were in the upper Edna Cintron. Was

which clearly were cut before it she whin was waving from9/ii
9/11 happened.                     thee North Tower hole/

Pieces of the *wreckage* of a *plane* that that never hit the
Pentagon but was a missile that went through to 3rd ring

Miss
ile trail of misssile NORAD saw it and gave the alert 6 minutes
before it hit the Pentgon but no planes were sent.

Pentagon hit by Russian missile  American journalist Serena no
Boeing 757 went through       Shims learned NGO vehicles

that hole was made by a Russian transported ISIS volunteers
missile,                          into Syria and was killed.

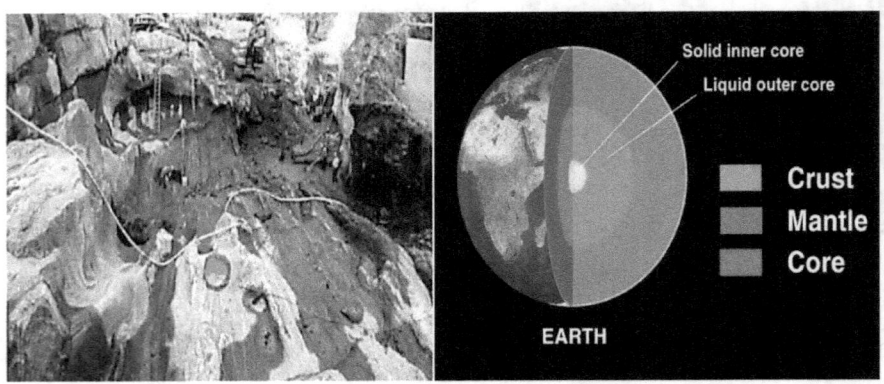

Crater 100 m (328') made by   Earth's molten iron mantel is mini-
nuke in basement of a      6,500 C., 1,000 C, hotter than
tower vaporizing bedrock.      the surface of the Sun.

Narcissist Caliph *El Baghdadi* Mossad   Mordecai Vanunu recruit
trained in Dimona to lead ISIS    Israeli Peace Activist

Israeli Planes attack USS
Liberty June 11, 1967

Russian jet landing at its base near
Latakia, Syria in October 2015

Israeli tanks invading Gaza kill
old men, women, and children.
Not exactly an even match/

Bilal Endogen buys stolen
Syrian and Iraqi stolen oil
from ISIS and supplies them

Israel night time bombing Israel
has a *right to defend itself.*

Grieving Gazan father holds
his dead son crying in agony

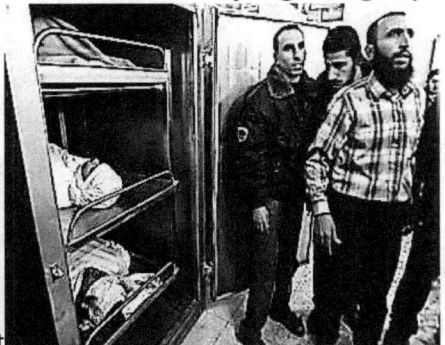

Unending line of small coffins  The Morgue always overflows

grows when Israel invades     when Israel invades.

Napalm, white phosphorus?  Looks. Jackie Sutton found hung
Like the *dustification* of twin tower  the women's restroom at
Ataturk International

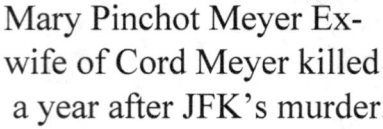

Mary Pinchot Meyer Ex-     Israel blocks cement imports to
wife of Cord Meyer killed   Gaza leading to more misery as
a year after JFK's murder.  well as sorrow over their dead.

Israel shell explodes in mid air  Israeli artillery shelling a hill
sending steel needles on Gaza. *Isarel's right to defend itself.*

## 9/11 The World's Biggest Insurance Scam

Larry Silverstein built the six building World Trade Center in 1970-72 for 90 million dollars on land owned by the Port Authority of New York and New Jersey.    PANYNJ controlled the complex until Larry and Frank Lowy acquired a 99 year lease on the complex on July 26, 2001 just 47 days before 9/11.  Larry leased 10.6 s square feet of office space but renters were few as there was a long wait for elevators to take workers to their offices.  Frank Lowy got 427,000 square feet of the retail *shopping mall* in the WTC basement.

In 1994, New Jersey Governor, Christine Todd Whitman, appointed Lewis Eisenberg Chairman of PANYNJ and it is he who stipulated that Larry Silverstein must insure rents for 99 years when the estimated economic life for a high rise is 99 years and thirty years had already passed.

In May 2001 Larry took over operation of the WTC after borrowing $726 million from CMAC (Commercial .Mortgage Corporation of General Motors).  He converted his loan into *securities* and sold his loan to *pension funds.*  But these loans had no collateral and were based on thin air.  Larry asked for and got $98 million back from the PANYNJ and paid lobbyists in Washington and Albany to limit his liability to the victims and families of 9/11.  Although asbestos was found to be dangerous to the health of people 1972, PANYNJ might have been forced to tear down the towers piece by piece. Thus 9/11 was not only an *act of God* and a blessing for PANYNJ that would have had people in New York City and adjacent New Jersey suing PANYNJ for endangering their health. All the insurance money to PANYNJ and Larry, zip for victims.

Liability for *loss of life* of passengers on any passenger plane is limited to 1.5 billion dollars. Was any compensation paid? The two flights from Logan landed at Cleveland, Ohio. and in 2003 an air controller at Cleveland International Airport identified one of the 9/11 planes from its numbers. CNN, *USA Today, The Boston Globe*, and the *Guardia*n in London published a partial lists of Passengers. But as no Arab names ever appeared on passenger, lists who provided the lists?

## The 1993 World Trade Center Bombing

Before the Bosnian War, the CIA had asked British MI6 for permission to use the Muslim Brotherhood to recruit a Muslim Foreign Legion (the *Mujahedeen)* to fight Russians in Afghanistan. MI6 gave them the blind Egyptian cleric, Sheik Omar Abdel Rahman and his followers. Sheik Rahman had been arrested and charged with the murder of Abdel Gamal Nasser in 1981 but as it was the Mossad that poisoned Nasser Sheik Rahman was freed. He was arrested again after Anwar Sadat was killed and as he was not involved was freed.

On October 2, 1984, the CIA flew blind cleric, Sheik Omar Rahman to Afghanistan to meet with Osama bin Laden. Sheik Rahman then recruited thousands of Muslim volunteers from Morocco to the Philippines and *the CIA flew then to southern Afghanistan, armed them and trained them* to fight the Russians and called them the *Mujahedeen*. The training camps were built and financed by the CIA, not Osama bin Laden. But when word of Serb atrocities reached the Mujahedeen in Afghanistan and James Baker's embargo on arms through the UN Security Council kept Bosnia, which as a member of the Nations it had a right to get Mujahedeen volunteered to fight in Bosnia and Osama bin Laden paid their way to Bosnia. Well, the CIA, Mossad, and State Department had not approved bin Laden's helping the Bosnians so the name for these fictional *terrorists* was *Al Qaeda* (The Base). That name came from the volunteers radioing those in Afghanistan with the call words, *Al Qaeda*. Osama bin Laden was then accused of exporting terrorism.

Chechnya had a right to secede from Russia with the break-up of the Soviet Union but but as the head of Russia's military, General Grachev, was not going to permit that and he was supporting the Serbs in Bosnia to keep Yugoslavia together so a Yugoslav Army would threaten NATO's right flank when Grachev moved to remake the Soviet Union, invade Ukraine, Poland, and the Balkans. Under Secretary of State, Strobe Talbot, swept the atrocities ordered by Yeltsin to support him as Russia's first president and Bosnians were accused of trying to create and Islamic Republic in Europe.

As Osama bin Laden had acted on his own, the CIA called him a terrorist where previously it lauded him as a hero. Osama bin Laden had a bank account in the BCCI and Ollie North used BCCI to his drug money that used to finance the Contras in Nicaragua. So the Southern District Court of New York was use to destroy the BCCI, Osama bin Laden's bank account, and Middle Eastern accounts that went to enrich the CIA's finances. That's how New York State's Attorney, Rudy Giuliani, qualified to be mayor of New York City, Strobe Talbot, became head Brookings Institution think tank, and blind Egyptian cleric, Sheik Omar Rahman, was set up as the mastermind of 1993 bombing of the North Tower to show Muslims as terrorists that wanted to hurt America. Bill Clinton was relieved of pressure to remove the arms embargo on Bosnia so it could buy weapons to defend their country from Serb aggression. It was the arms supplied Bosnia by Iran but that is another story as Iran also supplied Croatians with weapons that crushed the Croatian Serbs and over 100,000 fled to Serbia where Milosevic sent them to Kosovo to make trouble. And the lies to conceal the truth go on and on.

On February 26, 1993, the FBI said a van loaded with 1,336 pounds of nitrogen fertilizers exploded in the basement parking area of the North Tower. All around cars around this hole where the Econoline were vaporized or melted. Urea nitrogen fertilizer could never generated enough heat to melt cars. Only a mini-nuke can do that and Dimitri Khalezov has identified the 1993 bombing, twin tower bombings as well as the copy-cat bombing of Oklahoma City was done by mini-nukes. But based on the FBI's identification of a small piece of metal as coming from a white Econoline van that vanished in the explosion the Southern Federal Court of New York convicted and sent blind Muslims cleric, Sheik Rahman, to prison for life. He recruited Mujahedeen for the CIA to train in Afghanistan to fight the Russians that later were labeled Al Qaeda terrorists.

Eyal Ismail, a follower of Sheik Rahman, rented a white Econoline van and reported it stolen to the police and rental agency on February 25. On February 27, he went to the agency to recover his $400 deposit and this led to his arrest and Sheik Rahman and the others. Tried and found guilty, Sheik Rahman was sentenced to prison for life where he is today 20 years later.

N tman and the others. Tried and found guilty, Sheik Rahman was sentenced to prison for life where he is today 20 years litrogen fertilizer in a van used in the Oklahoma City bombing was to be blames on Muslims but it was Timothy McVeigh who did the bombing and he was recruited by the CIA with nine other men undergoing Special Forces Training to do government sanctioned assassinations and drug trafficking to finance covert operations.

William Rodriguez a Hero on 9/11

William Rodriguez was a janitor in the North Tower for 20 years. On the morning of September 11, 2001 he was late to work and instead of being in the top floor of the North Tower was in B2 basement at 08:12 when an explosion below his feet and almost lifted the North Tower out of the ground. The walls began cracking which the asbestos in the concrete was supposed to have prevented. William ran to pull Felipe David to safety. Felipe had skin dangling from his arms from burns he received which was like the victims of the Hiroshima bombing of 1945 who also had these kinds of burns which skin hanging from arms and torso. He then rescued two men from a flooded elevator shaft and seconds later an explosion he guessed was a plane hitting the tower was nothing compared to the explosion below his feet. While everyone was leaving the building, William stayed behind as he had been given a key that could unlock all the office doors after the 1993 bombing.

William went from floor to floor opening doors but the door to one unoccupied floor had noises like machinery rolling around he said he was afraid to open that door

On the 31st floor, he found a girl curled up and lying on the floor. He asked her why she hadn't left the building. She replied that this was her second day of work and didn't know where the exits were. William said, "All employees are to be shown all fire exits on their first day at work."

Firemen who climbed the stairs to the 31st floor with him, dropped their equipment and heavy boots to continue on up the stairs while William escorted the girl down the stairs and outside.

Just as he left the building an explosion blew out all the windows and glass was falling around him. The girl he had just saved was cut into by a shard of glass. When he turned to look behind him, there were bodies everywhere and the building began to collapse. He dove under a fire truck as debris and dust covered the fire truck. He prayed and was afraid he would die but was saved.

Invited to the White House, William was told that a movie about his experience was sure to come and his future would be rosy. But when he said the news media was in error and an investigation of 9/11 was needed, George W. Bush walked away from him and he was escorted out of the White House. He was found homeless and jobless living under a bridge and traveled the world to tell his story. In Venezuela, President Chavez gave him extra protection for he had experienced attempts by the Bush Administration and Cuban Exiles trying to overthrow his government where American oil companies had dominated Venezuela's oil exploitation under Venezuelan dictators for years. Where is William now?

## Five Dancing Israelis

The 9/11 Commission Report made by NIST omits the arrest of five Israeli Mossad on the afternoon of 9/11 which casts suspicion on its accuracy. Three of these men had cameras set on tripods aimed directly at the WTC twin towers from the New Jersey side long before the so-called planes hit the towers and were photographing and videotaping the burning WTC Tower. The New York Times reported on Thursday September 13, that five men giving each other hi-fives, jumping for joy in Liberty State Park and posing for pictures with the burning buildings in the background. It was so much like what the author saw in

Guangzhou, China at the wedding celebration at the time of 9/11 in New York City.

New Jersey residents were making angry calls to the police to report that middle-eastern men were videotaping the burning towers, dancing and giving each other hi- fives in a Jersey City park. "It looked like they knew what was going to happen when they were at Liberty State Park."

Later, other people witnessed them celebrating on a roof top in Weehawken (New Jersey), and later more reported seeing them "happy, you know ...Didn't look shocked to me." They saw these men making fun of burning twin towers and getting a kick out of photographing themselves in front of the horror. The FBI seized their cameras and developed the photos. One photo showed Sivan Kurzberg flicking a cigarette lighter with the burning towers behind him. According to ABC's 20/20, Sivan Kurzberg told the police, "We are Israelis. We are not your problem. Your problems are our problems. Palestinians are your problem."

An anonymous phone call a police dispatcher led authorities to close all of New York's bridges and tunnels. The caller told the 9-1-1 dispatcher that a group of Palestinians were mixing a bomb inside of a white van that was headed for the Holland Tunnel. Here's the transcript from NBC News:

Dispatcher: "Jersey City police."
Caller: Yes, we have a white van, 2 or 3 guys in there. They look like Palestinians and going around a building."
Dispatcher: "Where?"
Caller: "There's a minivan heading toward the Holland tunnel, I see the guy (going) by Newark Airport mixing some junk and he has those sheikh uniforms."
Dispatcher: "He has what?"

Caller: "He's dressed like an Arab. Palestinian."

Palestinians do not dress in galibeyahs but wear western style clothing, not *sheikh uniforms.* "Based on this phone call, police issue a "Be-on-the-Lookout" alert for a white mini-van heading for the city's bridges and tunnels from New Jersey.

When a van with that description was stopped just before crossing into New York, the suspicious *middle-easterners* turned out to be Israelis. The police and FBI field agents found maps with certain places of the city highlighted. There were box cutters which hijackers would use to stab or kill, foreign passports, and $4,700 in a sock. Police told the *Bergen Record* that bomb sniffing dogs reacted as if they had smelled explosives. Army Radio reported that three men were arrested and that the van *did contain tons of explosives.*

The *Jerusalem Post* reported on September 12, 2001 the day after 9/11, that: "American security services overnight stopped a *van packed with explosives* on the approach to George Washington Bridge. Authorities suspected **that terrorists** in the van intended to blow up the main crossing between New Jersey and New York."

The *New York Post* and New Jersey *Bergen Record* reported that a white van with Israelis was intercepted on a highway ramp near Route 3, leading to the Lincoln Tunnel. The *Israeli National News (Arutz Sheva)* and *Yediot America,* also reported a white van with Israelis was stopped on a ramp leading to the George Washington Bridge several miles north of the Lincoln Tunnel.

Were there two white vans? Eye-witnesses saw *Middle-Easterners* in a white van celebrating. It could explain how the New York Post and Steve Gordon; lawyer for the Israelis, said there were only three Israelis who were arrested, but it was later increased to five. The van with explosives would be parked on George Washington Bridge and the other van would pick up the driver and his helper and take them back to New Jersey. (Note that the vans kept away from the Holland Tunnel - which was where the police had been directed to go by the caller).

The Israelis were identified as *Mossad* by the CIA and FBI. They worked for Urban Moving Systems located in Weehawken, New Jersey. An American employee of the company told *The Record* of New Jersey that most workers *were Israelis* and *were joking* "and that bothered me."

A day later, Dominick Suter, the Israeli owner of Urban Moving System's, fled to Israel. Some Israelis were held in solitary for 40 days and released but the five Mossad were held for 71 days until Ehud Olmert, mayor of Jerusalem, called New York mayor, Rudy Giuliani to order the release of Five Dancing Israelis. No one was allowed to talk to them and all records are sealed. They were put on a plane and flown directly to Israel. Olmert went on to play a part in the destruction of a nuclear facility at El Kabir in Syria, September 7, 2007.

On return to Israel, three of *the art students* appeared on Israeli television. One denied they were laughing or happy on the morning of Sept. 11 and said, "The fact of the matter is we are coming from a country that experiences terror daily. Our purpose was to document the event." That is telling everyone they had prior knowledge of 9/11d why their cameras were aimed at Marsh McLennon offices in the North Tower where explosions ripped a huge hole in the building and a girl is seen waving from that hole. The *art students* knew how to *use explosives from their national service training in Israel.*

1. The Israelis were dancing and cheering the 9-11 attacks to celebrate the most successful and largest Black Ops ever pulled off and Americans with top security clearances aided them.

2. An accomplice calls a 9-1-1 police dispatcher to report Palestinians were making a bomb and heading for the Holland Tunnel in a white van but occupants weren't Muslims.

3. The Israeli Palestinian bombers then head for the George Washington Bridge (where they would ditch their time-bomb white van to escape with Urban Moving Systems co-workers).

4. Authorities immediately closed off ALL bridges and tunnels and that is how the Israelis were caught.

5. The *Justice Department* rounded up over 1,000 ethnic Arabs for minor immigration violations and put them in New York jails. But *immigration matters are not under the Justice Department showing this order came from* George W. Bush, The *Justice Department* said the five dancing Israelis were immigration violators caught in the same dragnet that had Arabs pulled in for questioning. After three months of detention, President Bush issued a gag order and their records are sealed.

When CIA Director, George Tenet was told a plane hit the North Tower, he said, "You know, this has bin Laden's fingerprints are all over it." But the FBI says they have no evidence linking Osama bin Laden to 9/11.

On February 13, 2001, the highly reputable journalist, Richard Sales, wrote that NSA had broken Osama bin Ladin's encrypted internet code, and there was nothing about 9/11 in the messages. This was seven months before 9-11 and there was nothing in the messages to indicate bin Laden was planning 911. Both the CIA and FBI have said *the attack on the WTC would have taken years to plan not just a few months..*

## Is Israel a Menace to America?

Abe Fortas successfully defended Lyndon Johnson against election fraud in his run for the Senate. LBJ appointed him to the Supreme Court in 1967 where Abe made it lawful for American Jews to have both American and Israeli citizenship. Australia then allowed its Jews to have dual citizenship. The Mossad duplicated Aussie Jews' passports to hide their Israeli nationality and registered as Australians in a Dubai Hotel where Mahmoud Al Mabhouh of the Hamas military was injected with muscle relaxant and then smothered with a pillow.

In 1954, Egyptian Jews recruited by Israeli Defense Minister Pinhas Lavon bombed American offices in Cairo to blame on Muslims were caught only when a bomb went off prematurely and severely burned one of their bombers.

In 1963 Israeli Prime Minister David ben Gurion gave the order to kill President John F. Kennedy which was carried out by three Jews in the CIA.

On June 11, 1967, Israeli air force planes attacked the USS Liberty and killed 32 American sailors and wounded 174.

On September 11, 2001, two trucks were in the tunnel under Building 4 and were being loaded with gold and when the South Tower was hit. The tunnel ceiling collapsed on the truck and the men took off. Mayor Giuliani reported $230 million in gold was recovered out of $1.08 billion in gold Comex and Federal Reserve stored in the WTC bombproof vaults so $850 million in gold was stolen.

The mainstream media ignored the 9/11 connection to Israel and focused on bin Laden. TV talk show hosts, *terrorism experts*, and news media people put out a continuous steady barrage of bin Laden was the leader of *Al Qaeda* exhorted Muslims to kill and terrorize. But Al Qeda were CIA trained Mujahedeen in Afghanistan and does not exist.

President George W. Bush said on TV that Muslims hate America's *freedoms.* But Muslim immigrants love America's freedoms as many don't have Freedom in their native country that is allied to America.

Shortly after 9/11 bin Laden was interviewed by a Pakistani and a BBC journalist and denied he had anything to do with 9/11. The denials were broadcast and published in Pakistan and Britain but not in the US. He said, "I was not involved in the 9/11 attacks in the United States nor did I have knowledge of the attacks. There exists a government within the government of the United States. The United States should try to trace the perpetrators of these attacks if they want to stop the present century from being a century of conflict between Islam and Christianity. That secret government must be asked as to who carried out the attacks. The American system is totally controlled by Jews, whose first priority is Israel, not the United States." All Yemeni Muslims are Jews that converted to Islam. Yemeni Jews that Immigrated to Israel are treated as inferior as they are darker, poorer, and less educated. As they have more children the hospitals will tell the mother their baby was born dead, and sell a baby to a couple who 'will give it a better live".

Osama bin Laden *has never* threatened the United States or *claimed* there he would more terrorism. Anyone publishing or broadcasting such statements is prevaricating.

A senior military officer commented that because of the visas and other documentation needed to get the supposed bombers into the United States it was inconceivable that Osama bin Laden had the capability to do 9/11

When CNN asked Rex Tomb, Chief of Investigative Publicity for the FBI why 9/11 is not mentioned on Bin Laden's *Most Wanted* web page. Tombs replied, "The reason 9/11 is not mentioned on Osama Bin Laden's *Most Wanted* page is because the FBI has no evidence connecting him to 9/11."

The only *evidence* is a barely audible fuzzy amateur video the Pentagon found in Afghanistan showing Osama bin Laden laughing with another man in a chamber in a mountain fortresses bin Laden built for the Mujahedeen *under direction of the CIA.* After 9/11 the translation of the conversation was changed to have bin Laden say, "and the walls will come crumbling down and thousands will die," then both men laugh.

On September 11, 2001, 09:40, a news media reported the Democratic Front for the Liberation of Palestine claimed responsibility for the attacks, but its leader, Qais Abu Leila, said the DFLP opposes "terror attacks on civilian targets, especially outside the occupied territories."

Carl Cameron of Fox News asked a US Official about the five dancing Israelis and their Mossad connection. The Official replied, "Evidence linking these (five) Israelis to 9/11 is classified. I cannot tell you about evidence that has been gathered. It's classified."

## Foreign News Media Reports on 9/11

"The Harbor Authorities of New York and New Jersey were the owners of the WTC. The twin towers, 412 meters high, were completed in in 1972, built by Larry Silverstein at a cost of only 37 million dollars. Since then, the towers have become a desirable address for corporate businesses. From completion of the buildings, office space has always been rented (not true) and the rents produced great returns for the owners. 430 companies from all over the world rented approximately 3.3 million square feet of office space in the WTC. More than 40,000 people were employed in the towers." Die Welt, 10/11/2001

"Three months before 9/11, the owners of the WTC leased the buildings to real estate tycoon Larry Silverstein. Since then Silverstein Properties Inc. took in the rent from the 430 tenants and other source income (i.e. tourist admission fees) and Silverstein paid a leasing fee to the owners: "Only three months before the attack Silverstein signed a rental contract for the WTC and agreed to pay a total of 3,2 billion dollars in installments over 99 years to the Port Authorities: 616 million was the initial payment, then 115 million dollars annually. The Port Authorities remained the owners of the WTC." --Die Welt, Berlin, October 11, 2001.

Despite not being the owner of the buildings, Larry Silverstein was the sole beneficiary of the insurance indemnity payments of more than 7 billion Dollars and signed the lease less than two months before 9/11. "Larry Silverstein, since July is the landlord of the towers, demands from the insurers 7,2 billion Dollars compensation," his spokesperson, Steve Solomon, said. ... The Port Authorities of New York and New Jersey, owners of the WTC, agreed with Silverstein's demand." --Die Welt, Berlin, Oct 10, 2001. (Called an *act of God*, Larry was indemnified and not PANYNJ the owners who said it all should go to Larry). "Within the next five years a new WTC could be erected on *'Ground Zero'*, Silverstein said after a meeting with the designated New York mayor, Michael Bloomberg." It will be called Freedom Center and will have two new subway lines built to La Guardia Airport and JFK International to offer duty free shopping for airline travelers. --Die Welt, Berlin, Nov 28, 2002.

"Saudi Arabia is now America's enemy, 'the core of evil' ... We Americans will annex your oil fields and your assets in America. The U.S. pushes Royal Saudi Arabia, just like Iraq, Iran, Sudan and Libya into an open membership club, called the *axis-powers of evil.'"--Die Welt, Berlin, Nov 23, 2002.

"70 year old Silverstein, by leasing the WTC only on July 7, 2001, made the 'dream of his life come true'." --Die Welt, Berlin, Nov 23, 2002.

"Daniel Liebeskind, the architect who designed the Jewish Museum Berlin was chosen to be the architect of Freedom

Center on Ground Zero, the NYT and NY1-Television reported Thursday." --Die Welt, Berlin, Feb 28, 2003.

September 8, 2001, saw "Trading with Sales-Options for United Airlines and American Airlines multiplied in the last three days before the attack." --Spiegel online, Hamburg, Sept. 19, 2001. (No evidence about 9/11 or how it happened is permitted in German courts of law)

All national and domestic security institutions and government agencies worldwide and the aviation industry must include terrorism in underwriting risk to conform with the Terrorist Risk Act of 2002. To offset the cost of terrorism insurance for the public to purchase, the federal government took some of the insurance industry's risk with The Terrorism Risk Insurance Act (TRIA) of 2002 to make terrorism risk available and affordable (based on the Muslims doing 9/11 and not Israel).

The Milliken Institute of Santa Monica, California, estimated 9/11 losses at $200 million. There was $35.6 billion for property, life, and liability not covered by insurance. The $850,000,000 gold bars stored in the WTC vaults by Comex and the Federal Reserve along with tons silver is nothing compared to the $2,300,000.000.000 trillion Dov Zakheim stole from the Pentagon for Israel. People believing the official report that planes flown by Arab Muslims brought down the twin towers and the Pentagon is called cognitive dissonance and hold to what they first heard is the correct version and what came after is a lie.

*Forbes financial review* on November 28, 2011 revealed that in November 2998, that Dr. Ben Bernanke of the Federal Reserve and Secretary of the Treasury, Paulson, went before Congress to say the US was just days away from a *financial meltdown* to ask for 16 trillion in *interest free loans* from November 2009 to November 28, 201. \

All as a result of George W. Bush allowed them to speculate in the derivatives market to make money lost by 9/11. Bush also said the derivatives markets *should not be regulated* and social security be privatized to invest in the US stock market. Goldman Sachs, Lehman Brothers, Morgan Stanley, Bank of America, Washington Mutual, held their hands out for a government bailout. Bank lending rules were changed. More home foreclosures, shopping malls, and small business died.

Larry Silverstein's insurers insured the rents at $3.5 billion for 99 years and Larry at the last moment slipped in a clause for each occurrence which he claimed for.

Larry Danielson, principal consultant at Deloitte said 9/11 forced carriers to look all lines of huge multi-risk loss and reinsurance coverage. Karen Pauli, senior analyst for the Tower Group, Needham, Massachusetts, said insurers had to introduce a new set of variable into the capabilities of CAT (Catastrophe) models) after 9/11.

Craig Lowenthal, was vice president and CIO of Hartford Financial Products and New York-based Integro Insurance Brokers ($300 million in private placement), said "One of the things we learned is where there is human intervention, it is error prone." Craig was at his desk on the 19th floor of Building 7 with a view of the north side of the North Tower.

Craig said Integro used a redundant network attached storage to back all information via LAN and WAN. But Dave Palermo VP of Sungard Capital's marketing (Wayne, Pa) said tapes are the Number-One method of backup and the first step of 100 steps insurers must take. All Hartford Financial documents were destroyed and Lowenthal had to go to clients, partners, and emails to rebuild their files.

Integro implemented IP telephony that can reroute all phones to a disaster recovery environment and allow employees to work anywhere. Mobility is essential for any industry to deal with catastrophes allowing employees to conduct business and communicate normally.

The New York Port Authority received 950 in insurance for the WTC built thirty years before for 90 million dollars for all seven buildings of which the Twin Towers cost only $37 million. Insurance fraud by companies with offices in the WTC has never been investigated. American and United Airlines never filed insurance claims for the loss of their planes.

The Southern District Court of New York ruled that the 9/11 World Trade Center attack was exclusion within an insurance policy that excluded insurance for property indemnified by a third-party. That barred the Port Authority of New York and New Jersey (PANYNJ) from coverage for certain WTC property damage.

The Southern District court of New York, the Port Authority of New York and New Jersey, No. 1:05 CV 05239, S.D. N.Y. ruled on February 22, 2008, PANYNJ could only claim for one of the six buildings and the commuter train terminal destroyed by the 9/11 attacks.

PANYNJ did receive a payment of $950 million for its losses but they allowed Larry to be the sole beneficiary. Larry became the owner of the land on which the WTC was built and is building the *Freedom Center* will be a *duty free shopping center* with new subways built directly to Kennedy International and La Guardia to whisk shoppers from the international airports directly to *Freedom Centers* so they can buy *duty free goods* on their shopping sprees to New York.

Insurers filed a declaratory judgment action in the Southern District of New York that PANYNJ's insurance did not insure losses to the Silverstein's Building 7 or train terminal. PANYNJ claimed an *exception that obligated* the insurers to supplement the leaseholder's insurance if it did not provide full coverage for losses. Insurers argued the exception applied only to other insurance and not to the indemnification. The court ruled the exclusion removed Silverstein Properties from coverage under the PANYNJ's insurance and PANYNJ was indemnified in respect to Silverstein Property and render the indemnification portion of the exclusion meaningless. PANYNJ was ever so careful to protect Larry. Why?

The Port Authority of New York and New Jersey adhered to traditional rules of contract interpretation and enforced the policy language as written, ruling that it cannot ignore clear and unambiguous policy terms as a full indemnification would render such terms meaningless.

## Pentagon Loss of Life from *Flight 77*

Milan Simonich's article in the Pittsburgh Post-Gazette reported that the missile hitting the Pentagon killed 34 of 65 civilian budget analysts, accountants, and bookkeepers going over the account books of Dov Zakheim on September 11, 2001, *their first day on that assignment.* Employed by Resource Services Washington, all were at their desks when the missile hit the Pentagon (six minutes after NORAD spotted the missile coming at the Pentagon at 2.5 mach speed. NORAD gave the alert and planes were in the air on a drill under V.P. Dick Cheney, *but no plane was sent to intercept.*

The 2.3 trillion dollars that disappeared under the Pentagon's comptroller who was also a dual Israeli-American citizen, Dov Zakheim reflects on Israel's currency, the shekel. The shekel was known as the world's worst currency but after 9/11 it now ranks with the Australian dollar, Canadian dollar, and Chinese Yuan as one of the top currencies.

The immediate disinformation was that Flight 77 was to have been directed at the Office of Naval Intelligence. The missile finally came to rest on the 3rd ring of the Office of Naval Intelligence and the whole section of the Pentagon was demolished through to the 3rd ring (p. 45).

When the investigation on the missing $2.3 trillion began to get hot, Dov resigned as comptroller got a job at Booz Allen, and Hamilton doing Strategic Studies. Booz Allen Hamilton also hired Edward Snowden who copied NSA files. Snowden is called a thief for only copying NSA files of our Government spying on all Americans but Dov Zakheim is a thief and is free.

. The FEMA Report stated the collapse of Building 7 "remains unknown at this time." On September 2, 2002, Larry Silverstein said on PBS Documentary *America Rebuilds*, "I remember getting a call from the Fire Marshal who said they were not sure they're gonna (sic)be able to control the fires. I said we've had such a terrible loss of life maybe the smartest thing would be to pull it."

Larry Silverstein has never issued a retraction of his order and photos taken of WTC 7 shows only small fires on two floors. The real reason to destroy Building 7 was that it had the backup files of the 2.3 trillion dollars Dov stole.

In February of 2002, Silverstein properties won an $861 million settlement from the Industrial Risk Insurers and the right to rebuild on the site of WTC 7. Silverstein Properties placed a tax evaluation of $386 million on the building. So the $861 was an increase of $475 million for destroying his own building and the insurance company and the court agreed?

The WTC construction manager said the twin towers were designed to withstand numerous plane crashes and could have withstood burning jet fuel.  FDNY chief of Safety reported that bombs were placed in both towers and *on the planes*.  A girl
is seen waving from the gaping hole in the North Tower (p. 430)
but no pieces of a plane were seen outside the building.

Colin Alexander calculated the trajectories of the plane flying past the South Tower are different from each other which
is proof  that the images of a plane were imposed on the videos and those who made the videos were involved in the conspiracy.

Several days before 9/11, the Israeli Embassy had many American  news media people in a meeting in the embassy and were. told to continually refer to 9/11 as an act committed by suicidal fanatical Muslim terrorists.

Over one hundred tons of *titanium alloyed steel*
*Jet engines disintegrated at the Pentagon along with aluminum fuselages and baggage were incinerated by the heat and explosions yet the DNA from 180 victim.*  Security tapes show no   plane hit the Pentagon.

All plane parts identified as coming from different planes and all the pieces collected as evidence have disappeared. Additionally there is no evidence of the plane touching the ground before hitting the steel reinforced concrete walls of the Pentagon at ground level.  All security tapes collected from businesses around the Pentagon show nothing.  One clip from

# Murder in the White House?

A Pentagon tape shows missile exhaust from the Russian Granit missile but no plane.  The explanation given is that the plane was going too fast to be recognized.

Bill Clinton's boyhood friend, Vincent Foster, was persuaded by Bill to be White House Legal Counsel.  Vince was brilliant, a highly successful lawyer making three times the salary of a White House Legal Consul but he took the job.  Vincent found the *travel agency in the Pentagon* was overbilling the Pentagon for billions and told Hillary and Bill about the overbilling but when nothing was done, Vincent was going to resign.

William Sessions, who was from Arkansas and was head of the FBI, on June 15, 1993, and Bill Clinton fired him. The next afternoon Clinton and appointed Louis Freeh as head of the FBI on recommendation of former President, George H.W. Bush.  Freeh had been Bandar Bush's attorney and George H W. Bush appointed hin as a Us District Court Judge for the SDNY in 1991 even though he had no prior experience as a judge.

Shortly after Freeh was appointed director of the FBI, Vincent Foster'sb was found in Ft. Marcy Park off Potomac Parkway.  Freeh was Bandar Bush's attorney and George H.W. Bush promoted him to be a US District Court Judge for the SDNY in 1991 even though Freeh had no prior experience as a judge.

His very first assignment was to investigate the death of Vincent Foster who was dead several hours before he was appointed director of the FBI.  The FBI under Freeh ruled Vincent's death a suicide.  Vincent was shot in the mouth with a

38 caliber revolver and the bullet exited the back of his skull and never found even though thorough search of the Civil War Park found thousands of objects from the time of the Civil War. Coroner noted that for this kind of wound there was little blood on the scene indicating he was killed somewhere else and body taken to Ft. Marcy Park.

Only one fingerprint was found on the gun in Vincent's hand and it wasn't Vincent's. The man who put Vincent's body in the body bag said he didn't need to wash his hands afterward as there was no blood on the ground or on the body which is unusual for a gunshot wound to the head. One bullet in the chamber yet no box of bullets was found in his room or office. Unusual as bullets are usually sold in boxes.

A blond hair as well as carpet fibers on his very expensive suit were never analyzed. All this indicates his body lay on a carpeted floor where he was killed and moved to where it was found but where was he killed?

One source states that Vincent's assistant. Linda Tripp went to work in the Pentagon for the travel office that was overbilling the Pentagon. Another says she went to work in the Pentagon's Public Relations. Monica Lewinsky confided to Linda that she was intimate with Bill and Linda Tripp taped their conversation exposing her sexual relations with President Clinton (read Wikipedia article on Linda Tripp).

Under Freeh, there was the Ruby Ridge killing, the burning of the Branch Dravidian complex, the Oklahoma City Bombing 1995. The Murray building bombing was a copycat of the 1993 Bombing of the WTC (urea fertilizer in a van) and to be blamed on Muslims but CIA agent Timothy McVeigh did it.

A Freedom of Information action was filed on the FBI to

make public the security tapes it seized around the Pentagon. A video clip shows the trail of the missile going into the Pentagon at near ground level (p. 44) but no plane hit the Pentagon.

Louis Freeh resigned May 4, 2001 and Thomas Pickering was made acting head of the FBI until Robert Muller took over on September 4, just a week before 9/11.

.

## How the Towers Were Brought Down

The towers were brought down in four phases. The first phase was to weaken the base so the building will come straight down and keep the upper part from toppling to one side. The weakening the central core was to disassociate all connections and the towers would collapse in a free fall.

The second phase is to initiate the collapse of the perimeter (outside) structure by attacking the four corners. There was red molten steel pouring out of the corners that was video-taped. That was to weaken the structure and keep it from falling to one side. An expert has to do the calculations. The possibility that a high rise building would come straight down in a free fall is zero. The buildings could not have collapsed like a house of cards, one floor dropping on the next floor as they were tied to a core of 169 steel elevator shafts.

Phase three is to hasten the collapse by disassociating the floor to perimeter column connections and the two vertical lines of spandrel plates at each corner of the tower. Continued attacks on the corners are to make sure the corners will be equally weak insuring a straight fall without panels falling outside the designated area of the collapse.

Phase four completes the collapse by attacking the remaining central core structure at lower levels and disassociating the horizontal bracing. Photos of the twin towers collapsing and flashes of light from explosions could not have been gas as no gas tanks or pipes were allowed.

The lower core structure must continue to sustain the building upright until the advanced stage.

+ Corners of the twin towers survived after frontal collapse.

+ Inward bowing of the perimeter walls and sagging floors?

+ Were there tilting movement of the upper sections?

+ Was there any bending of the upper section?

+ Early disintegration of the upper section.

+ Early downward movement of the antennae.

+There were ejections of dust and debris

+The flashes of light would be from explosives

+ Color and character of smoke emissions were different.

+There were unexplained molten metal ejections.

+ Behavior of the spires on top of the buildings

+ Flashes of light captured in photos were explosions.

## Power Down: Pre 9/11 WTC Security Blackouts

A few days after the World Trade Center was destroyed, the heightened security alert was lifted. Daros Coard, a 37 year old security guard at Tower One, said security detail had been working 12 hour shifts for two weeks before 9/11 due to threatening phone calls. On Thursday, September 6, bomb-sniffing dogs were removed from the entire WTC.

On the weekend of September 8 and 9, there was a power down condition in WTC Tower 2 (the South Tower) and no electricity for 36 hours for the floors above the 50th floor. Without electricity surveillance cameras and security locks were not operational and this would be the perfect time to plant demolition charges.  At this time, there were many engineers came and going.  Planning a demolition is a precise operation that calls for careful examination of the blueprints as a mistake can cause a building to collapse to one side.

## How Demolition Crews By-passed Security

Kevin R. Ryan of *Scoop Independent News* is the source of this information.   Kevin investigated who had the demolition expertise and access to explosive materials to do the demolition of WTC.  In the years preceding 9/11, tenants occupying the floors the planes struck had ordered structural modifications made to those same where the planes struck.

The demolition teams would need access keys to let them in the buildings to place and disguise the charges.   From whom d id they get the keys?  There were just five keys that were made to open all locked doors.  William Rodriguez had one, but the others were held by PANYNJ.

Were those hired by WTC to do electrical or construction upgrades and the five hundred million in upgrades by Kroll and other security companies make copies of keys?  Were those monitoring the security cameras involved?  Many in Silverstein Properties, PANYU, FBI, New York City police, and maintenance are most likely involved,

Lockheed Martin, Raytheon, General Dynamics (makes F-16 fighters and Gulfstream business jets (used by CIA for renditions and drug smuggling), Halliburton, and Science Applications International Corp (SAIC) - all have close ties to the Bush family corporate network, such as Dresser Industries (now

Halliburton), UBS (Union Bank of Switzerland), and Deutsche Bank. Its subsidiaries were benefitted from securities trades that one can only say were highly profitable trades no one else saw before 9/11.

Kevin Ryan's linking these tenant companies and the Bank of Credit and Commerce International (BCCI) and call it a terrorist logrolled bank for laundering drug money for the CIA and Oliver North is stretching it too far. Ollie opened an account for Juan Ramon Matta whose purchase of surplus flying boxcars only goes to show how the Mossad directed State Department diplomacy that affected Latin America and the economic subjugation of Latin America. Ollie North and General Secord persuaded ultra clean Ross Perot to co-sign on their accounts in order to hide their drug trafficking operation to finance the Contras in Honduras who committed atrocities. It was Rudy Giuliani's prosecution of BCCI that boosted him to the spotlight and put him in the position of being elected Mayor of NYC. But BCCI was destroyed because of two things. One was the money laundering it did for Ollie North's drug money laundering it did because the management throuth that it was acting on the orders of the Reagan White House and would get favored treatment. But it was also destroyed because it held the bank account of Osama bin Laden whose money had paid the way for Mujahedeen fighters to go to Bosnia to fight Serb Aggression and to Chechnya to defend Chechens who were being massacres by Boris Yeltsin. So the name *Al Qaeda* was created by the CIA to hide the fact they created the Mujahedeen and trained them in psychological warfare to fight Soviet Russians in Afghanistan developing Afghanistan's huge natural gas the Russians that would keep the Soviet Union alive. So during the Clinton Administration the Mujahedeen became *Al Qaeda,* and under the Bush Administration the plan was to invade Afghanistan and exploit its rich natural gas field.  Send it through Pakistan by Pakistan and  sell it to Japan and Pakistan.

The CIA claimed Al Qaeda had economic and financial ties with thirteen countries in 1999 and yet in 2009, some estimates of Al Qaeda's network was said to be 200 or 300.  Other sources have put Al Qaeda numbers as thousands with a presence in forty countries, excluding western countries.  The truth is Al Qeda is the Mujahedeen the CIA trained..

The June 1967 attack by Israel destroyed the ceasefire and was the beginning of Palestinians refugees as hundreds of thousands Palestinians fled from the West Bank to Jordan, and that became the origin of the PLO and Bashar Assad's father took over Syria and then next year Colonel Muamar Gadaffi overthrew King Idris.  Colonel Qaddafi, Hafez Assad, and Abdel Gamal Nasser of Egypt.  Nasser said that Gadaffi was a nice boy but naïve.  However, all agreed to an Arab Federation of the three countries.  Nasser came to the conclusion that the CIA would oppose anything he tried to do in developing Egypt to reduce unemployment at would cause instability.  So when Saddam Hussein was sent from Iraq to meet with Nasser and become the fourth nation in this confederation, Nasser introduced him to the CIA and they became his controllers of Saddam Hussein.  To nip this confederation in the bud, the Mossad poisoned Nasser in 1970 with poison in his massage oil even though Nasser was working with the CIA.  So who was more naïve?

From 1970 to 1978, Israel stirred up the Muslim Brotherhood what had heavy fighting between Assad and several Syrian cities.  Hafez Assad of Syria and Iraq's leader Al Bakr agreed to unite the two countries.  But Al Bakr stepped down *because of health problems* and the CIA selected Saddam Hussein to take his place as he was already cooperating with them.  Then Saddam was told that some 200 member of the Ba'ath Party were planning to overthrow him and he executed them which was what Israel wanted and they leaked this same kind of information to Ayatollah Khomeini to have him execute 200 Iranian Air Force pilots planning to overthrow him.

The CIA then whetted Saddam's ambition was to unite the Arab countries and replace the Saudis as head of the Arab World. By taking Khuzestan Province from Iran and annex it, the added oil reserves and oil production would be far larger than Saudi Arabia's production. The CIA arranged for 80,000 young Iranians to escape from Iran to Iraq and formed them into the Mujahedeen Khalk. The CIA financed them, trained them, supplied them with weapons, and turned them over to Saddam to fight Iran and Saddam then asserted that Shatt al Arab waterway was Iraq territory, but his goal was Iran's Khuzestan oil which with Iraq's oil was greater than Saudi Arabia and he would be head of the Arab World

The Mujahedeen the CIA calls the Al Qaeda, did not have any ability to do 9/11 or make mini-nukes that vaporized the bedrock on which the twin towers were built. But America, Israel, and NATO have mini-nukes. So who gave Mike Harrari the mini-nukes?. The CIA and NATO did *Gladio* Black Ops in Europe which were called terrorist operations to turn Europe against Communists. Like Betar, the,Zionist terrorists of Latvia killed thousands of ethnic Germans in Czechoslovakia in 1938 and thousands of ethnic Germans in Pomerania Poland in 1939 to get Hitler to invade Poland and Czechoslovakia. The Betar, Stern Gang, Irgun, Shin Bet, and Hagannah terrorists were integrated into the Mossad that has killed thousands.

A van filled with nitrate fertilizer could not blow a hole in the basement parking garage in the North Tower in 1993. The Oklahoma City bombing was also said to be done with urea nitrate fertilizer in a van parked across the street and was to b blames on Muslims. But Timothy McVeigh was caught by Oklahoma State Troopers and found guilty of killing 168 people and injuring 680. But McVeigh wrote his sister that the CIA hired him with nine other men to do *drug trafficking and government sanctioned assassinations.* The FBI took those letters and threatened his sister. So McVeigh was sentenced to death and while his lawyers were appealing the death sentence McVeigh said he wanted execution. He was given injection and no exhumation to verify that as his body was cremated.

Think more about the American Congressional and Presidential election of pro-Israel candidates from both parties that to prove Muslims killed 2,780 innocent people.

It cost only $90 million to build the WTC complex in 1972 and five hundred million dollars in security was installed in 1996 that didn't work on September 11, 2001. But there were billions more wasted on security as well. Besides all the benefits listed, Israel's Likud Party had a green light to label the Hamas government a terrorist organization and smash what was left of Gaza. Gazans chose Hamas in a democratic election as Hamas fought to protect Gazans. Israel bombed, strafed, **and** used white phosphorus on Gaza's civilians, cut off their food supplies, cut off fresh water, and claimed Hamas had fired rockets into Israel. When, and when the last defender of Gaza is dead, Israel will annex the remaining 30% of Gaza that UN apportioned as Gaza. The people of Gaza have suffered poverty, starvation, and voted for the Hamas government as it protected and looked after the people whose land was seized by Israel's annexations. Gazans have a *right to food, medicines, clothing, and building supplies* but its borders are sealed off by Israel and Egypt's military government. Unemployment is 49%. The Gaza flotilla bringing food, clothing, and medicines from Turkey was attacked in International waters by Israeli commandos dropped by

85

helicopters onto the skips and killed nine men. Forensic autopsies showed the Israeli commandos shot then in the back.

*Gazans dug tunnels under the Egy*ptian borders to bring in food and water to the people as Israel dug well to steal their stole the water from their springs. Israel said they were bringing in weapons like Jews did in the Warsaw Ghetto in WW II that wiped out two German divisions.

## PANYNJ and Its WTC Director

PANYNJ (Port Authority of New York and New Jersey) is inter-state agency behind the creation, construction, and management of the WTC complex until the last three months of its existence. PANYNJ was created in 1921 to "run the multi-jurisdictional commercial zones in the region." PANYNJ managed both the WTC and Newark Airport, involved in the terrorist attacks on 9/11. It also had offices on floors 3, 14, 19, 24, 28, and 31 of the North Tower.

PANYNJ's management had an executive director, a board of twelve commissioners - six of which were appointed by the Governor of New Jersey and the other six by the Governor of New York, and an executive director with recommendations from the governors.

Neil Levin was the Executive Director of PANYNJ six months before 9/11. He had been Superintendent of Banks for New York State and a general partner and vice president of Goldman Sachs. Geoffrey (Jeff) Wharton of Silverstein Properties finished his breakfast in the Windows on the World restaurant in the North Tower, where he always had breakfast, and was on the last elevator to descend at 8:44 AM. He met Neil who was waiting for someone he'd never met before.

Neil's wife, Christie Ferrer, called his office to ask his secretary: "Hi, I know you're crazed and don't want to bother you but the governor is looking for Neil and so am I, but no one can find him. Did you locate him?"

She replied, "I haven't got that information, but last I heard he was not in his office. Hold on a second," and she connected Christine to World Trade Centre director Alan Reiss. "I really don't know where he is," Reiss said.

"Do you know for a fact he wasn't in the office?" she asked.

"I don't know that for a fact," said Reiss.

Levin's body was found months later. He evidently didn't get the message received by the CIA at 6:45 am EST but evidently the CIA did not pass on the message not to go to WTC towers. His wife, Christine Ferrer, became Mayor Bloomberg's rep to the 9/11 Families and later appointed by New York Governor Pataki to PANYNJ's board of commissioners.

Alan Reiss was nicknamed the *mayor of the WTC* and lead person in the PANYNJ and intentionally deceptive after 9/11. In 1984, Reiss worked on building systems, engineering, and capital program, and eventually helped to secure major leases for tenants like Banker's Trust and Fuji Bank. In 1988, he was promoted to World Trade Center's supervising engineer, in charge of the complex's major systems and later promoted to Special Assistant to the Director. After the 1993 bombing of the North Tower, he managed the design and reconstruction and restored the WTC's infrastructure *'allowing the Twin Towers to reopen within a month.'*

Reiss' job required him to meet with top officials twice a day and he implemented many innovations and put forward a ten-year redevelopment program after the 1993 bombing of the North Tower that spent a *half a billion dollars (500 million dollars)* on upgrading the security of the whole WTC complex. The new

Security Command Center on the 22nd floor of the North Tower (WTC 1), and Operations Control Center in the B1 level of the South Tower (WTC2) *which became usel*ess as its offices in WTC 7 were destroyed.  Building 7 was evacuated and brought down that afternoon by the Fire Marshall at request of Larry Silverstein.

In 1996, Mayor Giuliani created the Office of Emergency Management (OEM) with offices in Building 7 *to unify operations between New York City's emergency responders.  It duplicated the NYPD's emergency center and absolutely useless on 9/11.  Was it a Giuliani payoff?*

Many PANYNJ managers were said to have trying to help others.  Alan Reiss survived and told the 9/11 Commission, "We were stunned when 2 WTC was *also* hit by a plane. The police officers and I rushed to the rear emergency exit and looked up at the tower and realized *we were at war*.  At the Port Authority Police's request, I returned to the vicinity of 6 World Trade Center to assess the condition of One World Trade Center with Captain Whitaker, commander of the Port Authority Police at the WTC.  Then the tower began to collapse and we were both enveloped by this churning black debris cloud.  As we ran north on West Street, it was darker than any burning building I have ever been in as a volunteer fireman, and it was next to impossible to breathe due to the debris in the air."

*People caught in the churning black debris cloud were covered in dust* but Alan Reiss *didn't have a speck of dust on him* when he appeared on television shortly afterward.  Alan said "If someone had told me the threat was a missile, I don't know what I could have done, let alone if someone had told me it was going to be a plane.  No one ever thought about a hijacked plane being rammed into a building."  Really?  There was the exercise drill that morning under the control of VP Dick Cheney to test the possibility of such an attack.

In 1984, Peter C. Goldmark, Executive Director of Port Authority of New York, created the Office of Special Planning (OSP) *that in 1985 called the WTC a most attractive terrorist target and speculated on a vehicle loaded with explosives in the basement of the basement parking area of a twin tower and* Science Applications International Corp, must have passed this information on the Sheik Omar Rahman who carried out the scenario Goldmark spent for much money on while he was under watchful eyes of the FBI.

Prior 1993 interview, Alan Reiss said that if an airliner was to crash into one of the towers, "The building's structure would still be there." In 2001, Reiss compared the energy of the planes' impacts to the *detonation of a tactical nuclear weapon* but said the buildings' construction may have saved lives." All engineers and architects consulted by PANYNJ after the twin towers were built said the buildings **could not be brought down.** However, Demolition Incorporated said it could be done with mini-nukes and thermite, and it was done like that

An estimated 100 fires around the world in high rise buildings every year and no high rise steel and steel reinforced concrete structure has ever collapsed. Alan Reiss said the impact of the planes hitting the towers was the equivalent of a tactical nuclear weapon. How he know about nuclear bombs?

Robert Boyle had twenty-two years in construction before joining PANYNJ. He was the Executive Director of PANYNJ from February 1997 until May 2001. After 9/11,
he returned to his former PANYNJ position temporarily but then became the President of **Empire Strategies**.

Robert R. Douglas was a Past Commissioners of the PANYNJ then chairman of Clearstream International that created hundreds of accounts in different banks to launder dirty drug money for government agencies    Two of those banks were
89

Bahrain International Bank and BCCI whose management accepted Ollie North's accounts.

Brian Michael Jenkins was a captain in the Green Berets in the Dominican Republic and later in Vietnam War (1966-1970). In 1999-2000, he was an advisor to L. Paul Bremer's National Commission on Terrorism. Jenkins led Kroll's analysis of future terrorist threats and predicted the possibility of "terrorists deliberately crashing a plane into the towers." After the 1993 bombing Jenkins said, *"We were the top security consultant in the nation,"* Jenkins worked on the five hundred million dollar WTC security upgrades after the 1993 bombing. From 1989 to1998, he was deputy chairman of Crisis Management for Kroll Associates.

Five hundred million dollars spent on security upgrades that were worthless on September 11, 2001 as was Giuliani's OEM. Giuliani rose to power on his prosecution of BCCI that laundered Ollie North's dirty money from drug trafficking that Juan Ramon Matta did for him.

Jenkins returned to RAND after his stint with Kroll and worked with Donald Rumsfeld, Condoleezza Rice, and Frank Carlucci of The Carlyle Group, Paul Kaminski of In-Q-Tel, and Francis Fukuyama (all of PNAC that called for a *New Pearl Harbor* to keep Americans on the alert and war on a continuous basis. The Carlyle Group profited greatly from its Iraq work. Russian Prime Minister, Vladimir Putin, stated that the Rand Corporation runs the ISIS out of the US Embassy in Ankara.

In 1934, Major General Smedley Butler disclosed that a group of wealthy men approached him to overthrow President Roosevelt. Butler headed the Marines that invaded Caribbean nations before WW I to collect money on defaulted loans that American bankers made with these nations. Smedley called himself a *racketeer* for he invaded and used men with guns to collect on the defaulted loans made to these poor countries.

Some of the wealthy men involved in the sttempt to overthrow Roosevelt in 1934 were John D. Rockefeller, Henry Ford, John and Allen Dulles, George Herbert Walker, and Prescott Bush, Senator and banker. Nelson Rockefeller always wanted to be the President of the United States and was made Vice President under Gerald Ford by a Congressional Act.

In 1933, Franklin Delano Roosevelt was controlled by Henry Morgenthau Jr. and guided Roosevelt into WW II to save England that held the Mandate for Palestine  His father, Henry Morgenthau Sr. guided Wilson into WW I, to get Palestne

Stanley Brezenoff was a Program Officer for the Ford Foundation before he became the First Deputy Mayor in New York City under Ed Koch. He took over PANYNJ from Stephan Berger (Dresser Industries). Brezenoff said of Kroll's work at upgrading of security at the WTC, "We have such confidence in (Kroll) I have followed every one of their recommendations."

In 2005, seven PANYNJ representatives contributed to the National Institute of Standards and Technology (NIST) official report... Their contribution looks to be intentionally falsified and it was Eglin Air Base that tested the steel to say it was inferior steel not fit for hi rise construction which is a felony. Larry Silverstein built the WTC complex in 1972 and should have been indicted.

Frank A. Di Martini, Manager of WTC Construction, said the towers could withstand multiple impacts from aircraft. He was the program manager for WTC security systems and died on 9/11.

In January 2001, Jerome Hauer, (Mayor Giuliani's director of the OEM from 1996 until February 2000) and was replaced at OEM by Richard Sheirer to become vice president and associate director at SAIC's *Center for Counterterrorism Technology and Analysis.* It is not clear if he spent only eleven months at SAIC, or if he continued to work there while also working for Kroll, but on 9/11, Jerome Hauer was "national security advisor with the National Institute of Health". Hauer was scrutinized over his links to the *anthrax poisoning scare* due to his positions at OEM, Kroll, and the anthrax mailings via SAIC and the interviews he did on 9/11.

Jerome Hauer told Peter Jennings on ABC, "Things were working as they should," Yet in his interview with Dan Rather on 9/11, he resolutely supported the official report of impact and fire-based failure of the steel structures which was contrary to what architects and engineers said. When Rather asked him "Would it have required the prior positioning of explosives to bring the buildings down?" Hauer quickly responded, *"No. The velocity of the plane had an impact on the building (and) planes filled with jet fuel...that the intense heat **probably** weakened the structure causing collapse."*

In May 2002, Jerome Hauer was appointed as Acting Assistant Secretary of the Office of Public Health Emergency Preparedness in the Bush Administration. In December 2004, he was present with military and Homeland Security representatives at the National Press Club Bio Defense Briefing Series along with George Lowell, Chief Scientific Officer at ID Biomedical Corporation a subsidiary of Glaxo Smith Kline that manufactures the H1N1 swine flu vaccine that CBA 60 Minutes in 1976 reported as unsafe and unnecessary. But in 2009 all US troops were vaccinated with H1N1. Apollo Management shipped bulk vaccine to processing facilities for CEVA. Apollo Management was founded by Leon Black, who was on the PANYNJ Port Security Task. Jeffrey Benjamin: and WTC tenants, Bankers Trust, Exco Resources, and Henry Silverman of Cendant, the PANYNJ, and A.B

Krongard's wife, Cheryl Gordon Krongard are involved in CEVA. In late 1998, CIA Director George Tenet and CIA executive director A.B. Krongard called on Augustine for advice on defense changes. Congressional investigators were suspicious of Augustine profited so much from the CIA funded In-Q-Tel he created. After 9/11 the hype about fanatical Muslims killing 3,000 Americans and the Congressional suspicions about Augustine evaporated.

In May of 2004, Kroll *was purchased by Marsh & McLennan for $1.9 billion.* Jules Kroll pocketed more than 100 million from the sale. Kroll's CEO Michael Cherkasky became the CEO of Marsh & McLennan and former CEO, L. Paul Bremer, persuaded Eliot Spitzer to drop fraud charges against Cherkasky for cheating clients in price fixing and bid rigging. Jeffery Greenberg resigned as Marsh & McLennan CEO. Remember, Kroll's management created the WTC security and Marsh and McLennon offices were on the floors identified as the floors hit by the first plane reportedly seen by Joseph Pfeifer (see p. 91). . AIG was one of the owners of Kroll as of 1993 along with Marsh, Lockheed Martin, and L-3 Communications (Loral Corp) which had its stock flagged for by the SEC a few weeks after 9/11 for *insider trading.* AIG's chairman, Maurice Greenberg said: "Opportunities for us are enormous. It's not just in the United States. *Interest rates are rising throughout the world and business looks quite good,*" yet they got hundreds of millions in the *bailout of trillions*.

President George W. Bush, Donald Rumsfeld and Condoleezza Rice all said, *no one* ever thought about a plane hitting the twin towers. However, Richard Clarke, Counterintelligence under Condoleezza Rice, sent a memo on January 25, requesting a high level review on Strategy for Eliminating the *Status of the Jihadist Networks of al Qaeda Status and Prospectus* in September 1998 Pol-Mil Plan for al Qaeda. The Delenda Plan to kill Osama bin Laden was for  ABC John Miller who gave bin Laden a battery that would guide a missile to kill him.  But Osama bin Laden learned about the true purpose and turned off his phone.

The NIST report showed how security people at the WTC went from one company to another or to a government position holding multiple jobs at same time like Dov Zakheim held four to six positions at the same time.    His grandfather was a Bolshevik/Communist yet the Pentagon gave him top  security clearances even through the American government had years of searching for Communists after WW II.

John Deutch was with SAIC investigating the 1993 bombing of the North Tower.   In 1994, he became Deputy Secretary of Defense.  In, 1995, he, replaced James Woolsey as Director of the CIA.  In 1997, Deutch, Louis Freeh, Brian Michael Jenkins, and retired Air Force General James Abrahamson (Securacom) recommended that all passenger planes have Global Positioning System installed and put the Federal Aviation Administration's Wide Area Augmentation System under the supervision of the Department of Defense.  Deutch had worked for Raytheon and Science Applications Corporation (SAIC).

Brian Michael Jenkins and Securacom's General Abrahamson recommended guidance devices be installed on commercial aircraft that could take control of a hijacked plane. But *since 1970*, Pike's Peak NORAD's military facility had technology that could take over the controls of 4,000 passenger planes in the US air space at any given time and land them.

Louis Freeh resigned from the FBI on May 4, 2001 and *started a business that has profited well from the newly created Homeland Security industry.*  Freeh also became the personal attorney for Saudi Arabia's Prince Bandar.

Jules B. Kroll is founder of Kroll and a graduate of Cornell University. Admitted to the New York Bar in 1967, and was an assistant District Attorney for New York City. He was also a member of the secret society *Quill and Dagger* (as were former Deputy Secretary of Defense Paul Wolfowitz, National Security Advisors Sandy Berger, Stephen Hadley, and Stephen Friedman who became George W. Bush's top economic advisor. Friedman was also a member of the *Brookings Institution, Bilderberg group, the Foreign Intelligence Advisory Board,* and on the board at *In-Q-Tel.*

Neither the 9/11 Commission nor NIST (National Institute of Standards and Technology) mention Kroll in their reports. Kroll had a role in the security plan for the WTC and boasted it was "*the world's leading risk consulting company.*"

Kroll worked for corporations that wanted to find fraud or secrets within an organization such as a corporation threatening another corporation with hostile takeover. Kroll would find embarrassing secrets about the takeover company's executives that would prevent a takeover. Kroll hired former CIA spies, FBI agents, and prosecutors. Some made twice the salary they made in the public sector while investigating corporate and personal misdeeds. One of Kroll's important cases involved the derivatives fraud at Daiwa Securities. Kroll was called "The CIA of Wall Street" after tracing and recovering assets of Ferdinand Marcos for the Philippines, Jean-Claude Duvalier for Haiti, Saddam Hussein assets for the Russian government and for the Federal Deposit Insurance Corporation (FDIC) during the Savings and Loan scandal.

Kroll Inc. leaders have included Robert J. MacGuire, who was President and COO of the company from 1989 to 1997, and was formerly the NYC Police commissioner. Kroll board member, then Raymond E. Mabus, a lifetime Democrat, and Governor of Mississippi from 1988-92, who Bill Clinton appointed US

Ambassador to Saudi Arabia from 1994 until 1996. His position at Kroll is in no way connected to 9/11.

James R. Bucknam, Kroll's executive vice president for operations, was senior adviser to (Bandar Bush's attorney) Louis Freeh. Bucknam was an assistant United States Attorney in the Southern District of New York (AUSA SDNY) from 1987 to 1993, at the same time Rudy Giuliani was the US District Attorney for SDNY, investigating BCCI.

By 1991, Jules Kroll was called an expert with knowledge of how BCCI moved and hid money all over the world. But BCCI had nothing to do with 9/11. It was that Osama bin Laden's bank account with BCCI and he paid for plane flights the Mujahedeen took to fly to Bosnia and Chechnya which the CIA and Mossad called *Al Qeda*. The FBI and CIA destroyed BCCI was that it was the *only Middle East bank in Europe* with billions of dollars from Iranian and Gulf businessmen. Freezing bank accounts is used to destabilize nations along with sanctions on trade, arms embargoes. People deemed anti-American are put on the *no-fly list*, and George W. Bush added kidnappings, torture, arrest and detention of Americans under the Patriot Act, which was Unconstitutional).

In 1993, American International Group (AIG), was led by Maurice Greenburg, who owned 23% of Kroll. His son Jeffrey, is a member of Brookings Institution and Trilateral Commission. He left AIG two years later for Marsh & McLennan. Jeffrey rose to CEO of Marsh when non-existent *American Airlines Flight 11* struck the floors of the North Tower exactly on floor Marsh & McLennon occupied. Was Jeffrey in his office?

August 1996, after TWA Flight 800 to Paris was split into by missile fired from the USS Normandy (to see if the missile could tell *friendly from unfriendly planes*), Kroll began publishing a monthly reviews on air travel safety but Jacques Chirac never forgave Clinton for allowing the US Navy to test a missile in air corridor full of passenger. Pierre Salinger (JFK's press secretary) tried to expose Clinton but couldn't.

On the morning of September 11, 2001, New York City's Fire Department Battalion Chief, Joseph Pfeifer, was investigating a gas leak close to the WTC. He said he heard the noise of a jet plane and looked up to see Flight 11 hit the North Tower and explode. He said the lobby of the North Tower where it "looked like a plane hit it with broken glass everywhere," And again he heard the sound of a second plane and saw it hit the South Tower, which is questionable.

Colin Alexander has put together evidence that no plane hit the South Tower and images of a plane flying into the South Tower were digitally inserted on the video tapes.

## Electronic Systems Associates and E.J. Electric

The security plan at the WTC complex involved more than changes in infrastructure. Ongoing security also included the use of PANYNJ police (PAPD), as well as contract security personnel. In addition to those managed by the PANYNJ, some tenants had their own security contractors, and several of them died on 9/11. The PANYNJ itself lost a total of 84 employees, including 37 police officers, and its police superintendent.

When comes to placement of explosives for demolition in the Twin Towers, it is important to know the modifications and construction work that was performed in the years before 9/11. The police were just doing their jobs and those who died did not benefit from the terrorist attacks. The Port Authority Police Department (PAPD) were the first respondents like the Fire Department of New York (FDNY) and they lost many men because the buildings to be wired for demolition.

After the 1993 bombing, the new security plan for the WTC required years of work and involved many contractors. But there were four security contractors that led this effort. The contractors responsible for installation of the system throughout much of the towers were Electronic Systems Associates and E.J. Electric Installation, both of New York. The security of the basement parking garages should have been the most important area after the bombing and yet it was circumvented. Charles Finkel and Ensec's director Terry McAuliffe installed in the basements of the towers that didn't secure the buildings or were by-passed with the knowledge of how to py-pass them.

The evens of 9/11 show favoritism in handing out lucrative contracts to favored groups and sloppy work performed. The explosions that took place in basement level B5 and B6 just a minute before the 'plane' hit the North Tower had no security cameras. The two major companies overhauling the WTC security systems after the 1993 bombing were Ensec and Securacom,

Ensec International was founded in Brazil in 1983, by Charles N. Finkel of Brazil who was also its CEO and export sales executive for Ensec's Florida subsidiary. Ensec had an office on the 33rd floor of the North Tower (WTC 1). Ensec was the *sole partner of Mayfair Limited Partnership* which was a limited was a joint venture that included Apollo Real Estate Advisors formed in April 1993 by Leon Black as an affiliate of Apollo Manager formed in 1990 and *financed by US$159 investor funds.*

Ensec manufactured the tanks and military vehicles for *Operation Desert Storm* the first invasion of Iraq. It was involved in upgrading the WTC security after the 1993 bombing and boasted: "We are very proud that Ensec has been selected as the supplier of a major component of this highly sophisticated system." It designs, develops, assembles, sells, installs and services security systems for large commercial or governmental facilities ranging from single function installations to highly integrated security systems." Ensec installed the security systems in the basement levels of the Twin Towers but there were no security cameras installed in B5 and B6 and it was easy to place mini-nuclear devices to destroy the base of the steel structure for a hi-rise demolition and the heat from the mini-nukes vaporized the bedrock.

Ensec manufactured and installed the *access control system* that included proprietary software, proximity card readers and vehicle identification tags for all registered vehicles and cameras in computer areas, and visitor areas.

Electronic Systems Associates (ESA) was a division of Syska & Hennessey (S&H), an engineering firm that did structural engineering for the United Nations building, WTC building 7, and King Saud University in Saudi Arabia. S&H had offices in Saudi Arabia, and in Iran before Khomeini. In 1997, it was the prime contractor to the U.S. Army Corps of Engineers. Employees of ESA reported having worked on projects for clients such as The Pentagon, The Union Bank of Switzerland, March & McLennan and Hearst Publishing.

E.J. Electric Installation Company was founded in 1899 and hired Jacques Mann in 1912. Jacques' son, J. Robert Mann entered Yale University where he joined the *Skull and Bones* secret society that included Jonathan J. Bush, of Riggs National, George Herbert Walker III (a cousin to George H.W. Bush) the chairman of Stifel Financial Corporation (and Ambassador to Hungary) in 2003 and won a Yale Science and Engineering and "Award for Meritorious Service to Yale University." Robert graduated in 1951 and became chairman and CEO of E.J. Electric Installation.

J. Robert Mann son, Anthony (Tony) Mann, joined E.J. Electric in 1986. Some of E.J. Electric's past jobs included the Rockefeller Center, The Merrill Lynch's primary data center at the World Financial Center, AT&T World Headquarters in Manhattan, American Airlines, and the installation and maintenance of all voice, data, audio, video, satellite, security, and fire safety systems for U.N. buildings in New York and New York City's 9/11 Police Command Center, Delta Airlines, Tower Airlines, and British Airways at JFK Airport. The U.N. building in New York has been the site of electronic espionage which is not surprising as the World Bank in Washington has hidden cameras in all offices and records all telephone conversations on the biggest IBM computers in a huge room on the top floor in Washington. The author's office was the only office on the top floor. One day when the door was open he saw it all. When the author spoke to personnel about the spying she said US laws did not apply to international organizations.

E.J. Electric did work for United Airlines, Silverstein Properties, Saudi Arabian Airlines, and the Federal Reserve Bank and did the maintenance at the WTC after doing the security wiring, fire-alarm systems and back-up power.

On September 11th 2001, E.J. Electric had an office in the South Tower with nine electricians on-site, including a man on top of the tower before the first strike. E.J. also had a team in the basement. Tony Mann, President of E.J. Electric said that when the South Tower was hit by a plane all of his electricians ran out of the building and kept going. "A lot of them didn't stop until they got to Central Park."

E.J. Electric worked with the FBI in its investigation of 9/11. "We did the whole security system down there and provided our entire database to the FBI," Robert Mann said. "They had ten FBI agents at our subcontractor's office in California to download the database, which included everyone that worked and photographs were also included."

Lockheed Martin IMS subcontracted its portion of its PANYNJ contract to Ensec in November of 1996. Lockheed's CEO, Norman Augustine, worked with many people mentioned in the review of tower tenants such as Toru Hashimoto of Fuji bank, William Clark of the Washington Group, and Stephen Friedman of Marsh & McLennan. Augustine was a board member of Riggs National, the BCCI-linked banking firm.

Electronic Systems Associates (ESA) was a division of Syska & Hennessey (S&H) an engineering firm whose largest projects in

100

the past included serving as structural engineers for the United Nations building, WTC building 7, King Saud University, and S&H offices in Saudi Arabia,. Before the Shah was ousted they had offices in Iran. In 1997, the firm was retained as the prime contractor to the U.S. Army Corps of Engineers. Employees of ESA reported having worked on projects for clients such as The Pentagon, The Union Bank of Switzerland, Marsh & McLennan and Hearst Publishing.

E.J. Electric Installation Company was founded in 1899. It hired Jacques Mann in 1912. His son J. Robert Mann, a 1951 **graduate of Yale**, was a winner of the Yale Science and Engineering Award for Meritorious Service to Yale. Robert joined the Yale *Skull and Bones* secret society that included Jonathan J. Bush (Riggs National bank where Norman Augustine is a director), George Herbert Walker III, (Chairman of Stifel Financial Corporation) and ambassador to Hungary in 2003. J. Robert Mann became chairman and CEO of E.J. Electric and his son, Anthony (Tony) Mann joined the company in 1986. J. Robert Mann was senior advisor for the investment company Frontier Group working with Frank Carlucci of the Carlyle Group but retired in 1997.

Norman Augustine served as president of Lockheed in 1995 and became its CEO later that year and then the CEO of Lockheed when a subcontract was awarded to Ensec. Ensec Lockheed has benefitted greatly from the War on Terror and Lockheed Martin became more than just a giant military contractor when it bought Datacom Inc. and changed it into Information Management Services.

IMS works in 44 states and several foreign countries and does everything from collecting parking fines, running welfare to work job training programs, and training airport security workers. One airport security worker stole $800,000 in valuables from checked in luggage. IMS also trained interrogators that interrogated detainees at Guantanamo, at Bagram in Iraq,

Afghanistan, Morocco and other countries. Under pressure from Wall Street to concentrate on its core business, Lockheed Martin sold IMS in 2001, which had deals with the IRS, the Census Bureau, and U.S. Postal Service.

Norman Augustine was in US policy discussions with the need for changes to defense and space program spending. In 1990, he chaired the Advisory Committee that included George H.W. Bush's Vice President, Dan Quayle, (a PNAC signatory) to the U.S. Space Programs known as the Augustine Commission. In 1990, Augustine co-wrote The Defense Revolution, with WTC 2 impact zone tenant Joseph Kasputys contributed to the 1996 book Defense Conversion.

In late 1996, Lockheed subcontracted its PANYNJ work to Ensec and added two new directors, Raymond E. List of ICF Kaiser Engineer and Terence R. McAullife. In 2001, he became Chairman of the Democratic National Committee (DNC) and linked to the Teamsters Union made a $1.35 million gift to the Clintons. McAuliffe was also involved in a lawsuit where Loral Space gave secrets on satellite and intercontinental ballistic missile programs to China. Prescott Bush Jr. (George W. Bush cousin worked with Hughes and got satellite deals with China.

McAuliffe in the Loral Space scandal was charged with agreeing "to sell seats on taxpayer-financed foreign trade missions and other government services in exchange for campaign contributions to the DNC" and "in securing other favorable treatment from the Clinton Administration for Loral." Among those selected in the high-profile Commerce Department trade mission was Bernard L. Schwartz, (billionaire CEO of Loral and *Globalsta*r) became the biggest contributor of the Democratic National Committee..

The CEOs of Hughes, Loral, and Lockheed co-wrote a letter to President Clinton in October 1995, asking him to "transfer all responsibility for commercial satellite export licensing to the Commerce Department," But Hughes, Loral, and

Lockheed was found guilty for giving advanced missile technology to China and paid enormous fines.

In April, 1996, Loral was close to bankruptcy when Lockheed Martin bought it for $9.1 billion. PNAC co-founder, Francis Fukuyama, is a Bernard L. Schwartz Professor of International Political Economy at Johns Hopkins University.

McAuliffe purchased $100,000 of Global Crossing's stock before it went public and cashed the stock out several years later for $18 million, which was a tidy profit.

Richard Perle was a lobbyist for Global Crossing, which is a partner Chinese Hutchison Whampoa, which is an *arm of the People's Liberation Army* of China.

Li Ka-Shing owned Hutchinson and invested in firms owned by Winston Partners, and Winston's cofounder is George W's brother, Neil Bush, was hired as a consultant.

Between McAuliffe's Democrat National Committee role and campaign chairman for Hillary Clinton's Presidential run he worked as Vice-Chairman of Carret investments. He was hired at Carret by Alan Quasha, who bailed out George W. Bush's oil company in 1986, by putting it into his company, Harken Energy. Harken Energy had connections to BCCI. Businesses crossing party lines involve businessmen who are shady. That is what advisors on the boards of companies look for to avoid pitfalls.

Hassan Nemazee esd a partner and investor in Harken, a founder of the Iranian-American PAC, a member of the Council on Foreign Relations, and connected to the Brookings Institution that is connected to RAND which was one of the corporations that approached Major General Smedley Butler of the Marine Corps, to overthrow President. Roosevelt in 1934. But General Butler went and told on them, but no one was ever charged with treason or an attempt to subvert the Constitution.

## Securacom and Security Issues

Burns & Roe changed its name to Securacom Inc. in October 1992, declaring it was a "large-scale security and facility management for businesses and government." The company's first job was with WTC in 1993, following the bombing and was awarded a larger WTC contract in 1996. In 1997 Securacom's name was changed to Stratesec over a lawsuit on trademark infringement. For the simplicity the name Securacom will be used instead of Securasec. Securacom tried to bury the plaintiff, Ron Libengood, and take everything he had by filing frivolous arguments in multiple jurisdictions.

In 1997, Securacom worked on projects for Amtrak, Xerox, and "a joint venture agreement with Ahmad N. Al Binali & Sons Co., a large Saudi Arabian engineering and construction company, "to develop and conduct business in the Kingdom of Saudi Arabia." Amtrak is a federal government entity to provide railway passenger service all over America. The head of Amtrak has always been a Jew.

Jews have been forbidden to enter Saudi Arabia for years. The author does not know if this was because a Jewess, Zindel Harib invited the Prophet Mohammed to dine in her home and served him poisoned the food from which he died.

Many Jews do work in Saudi Arabia and it was Chaim Weisman head of Britain's Jews who requested Winston Churchill to make King Saud head of the Arab World in January 1945. Churchill's American grandfather, Leonard Jerome was a Jew whose wealth came from being the largest stockholder of the *New York Times* and Churchill halted the British financing of the White Russian Army that was defeating the Red Army in 1920 by saying the White Army was persecuting Jews. Not true as the White Army was pursuing Communists who were Jews.

Jews are supposed to be forbidden to enter Saudi Arabia which is not true as many Jews work in Saudi Arabia. Saudi

Arabia was made head of the Arab World by the Jews as Chaim Weisman head of Britain's Jews asked Winston Churchill make the Saudis the head of the Arab World in January 1945 and Churchill's American grandfather, Leonard Jerome was a Jew whose wealth came from being the largest stockholder of the *New York Times*.

The prohibition against Jews to enter Saudi Arabia was handled by the commercial attaches in American Commercial Attachees in American Embassies that gave American-Jewish businessmen *Christian Baptismal Certificates* to enter Saudi Arabia to do business. But another prerequisite to enter Suadi Arabia is that they also had to have a letter of invitation from a Saudi company to vouch for them.

Jews in the Department of Commerce hated the Saudis for demeaning them. But Securacom, which has many Jews in its management, has a joint venture with Ahmad N. Al Binali & Sons Co. so this prohibition is just lip service.

The author knew personally a Jewish girl who was employed to teach English to Saudi Air Force pilots. But *bending the rules* works in Saudi Arabia and the Emirates.

Israelis still have this great antipathy toward Saudi Arabia and Israeli agents are loud and strong in saying the Saudis are trying to create a Caliphate and are supporting ISIS in Syria and Iraq. Crazy, as a Caliph cannot inherit the title but has to be elected. Shiites say a Caliph must be a descendant of the Prophet Mohammed while Ismaili Muslim's differ from all others as their Agha Khan, a direct descendant of the Prophet, is the *hidden Imam*. However, the majority of Ismailis are Losana Hindus (Khojas) the Agha Khan converted to Ismaili Islam in Bombay after the British rescued him from Iran where the Shah wanted to kill him for helping the British invade Iran.

Securacom provided the security services for United Airlines at Dulles Airport where United Flight 77 that *hit the Pentagon*. The United Flight and the American Airlines Flights were non-scheduled flights. United Flight 77 went off the radar after leaving Dulles but another radar tracked to the Ohio-Kentucky-West Virginia area. The American Airlines planes were said to have landed at Cleveland Airport.

Securacom went public on September 11, 1997, its Initial Public Offering prospectus featured photos of the World Trade Center, Dulles airport, United Airlines, and Los Alamos National Laboratories. On its board of directors were retired Air Force General James Abrahamson, Marvin Bush ( brother of George W.), Wirt Walker III (nephew of George H.W. Bush), Charles Archer (former Assistant Director in charge of FBI's Criminal Justice Information Services Division under Louis Freeh), and Yousef Saud Al Sabah, of the Kuwaiti royal family.

Kuwait had entered disputed border areas of Iraq/Kuwait during the Iran-Iraq War and drilled for oil and used some of the money to support Iraq. Saddam considered this a breach of their agreement and consulted US Ambassador April Gillespie. When she didn't reply to his intention to invade Kuwait, Saddam thought it meant the Bush Administration accepted it and he invaded Kuwait. To way US public opinion and support for the former head of the CIA, President George H.W. Bush's 1991 invasion of Iraq, it was necessary to make Saddam a monster. So a fifteen year old Kuwaiti girl, Nayrih, testified to the House *Human Rights Caucus* that "Iraqi soldiers came into a hospital with guns and went to the room where babies were in incubators. They took the babies out of the incubators and left them on the cold floor to die." Nayrih was not any Kuwaiti girl but member of the Kuwaiti royal family, the daughter of Saud Nasir al Sabah the Kuwaiti ambassador to the US. She was coached to lie by the Public Relations firm of Hill & Knowlton.

Another member of the Kuwaiti royal family was Yousef Saud Al Sabah, chairman of the Kuwait-American Corporation (KuwAm), which, between 1993 and 1999 with controlling shares of Securacom. How could Securacom whose majority shareholder, a non-American, do the security for sensitive US facilities? Easy, Israel was given that privilege as well. And look at how many dual-israeli citizens are elected to Congress and some Jewish Americans who had connections with Communism get top security clearances when America was in conflict with Communist countries such as Viet Nam and taken adversarial positions in Cuba, Angola, and Latin America?

KuwAm's computer was linked to those of Securacom so some of KuwAm's employees could have fed information to Israeli agents. Wirt Walker III was linked to Al Sabah's company called Special Situation Investment Holdings (SSIH) that owned several other companies (and Commander Aircraft and Aviation General).

Retired General James Abrahamson was President of Hughes Aircraft, and a director of Turkey's Global Group. In December 1977, General Abrahamson, a former NASA administrator under the Reagan Space Defense Initiative (SDI) became a director of Securacom. He also served on a White House Commission on Aviation Security, and from 1984 to 1989; he collaborated with Deputy Assistant Secretary of Defense Frank Gaffney and Assistant Secretary of Defense Richard Perle, both PNAC members.

Abrahamson was also President of Hughes Aircraft from 1989 to 1992, when Prescott Bush Jr. was lobbying his brother, President George H.W. Bush to lift sanctions on the Chinese government. Abrahamson now serves on the board at GeoEye that spies on everyone on earth by way of satellites. He is also a co-founder of a company called Crescent Investment Management (Crescent), with Mansoor Ijaz.

Abrahamson's business partner, Mansoor Ijaz, is CEO of Crescent, and on the advisory board Maurice Sonnenberg a former CIA Director James Woolsey are PNAC signatories. Ijaz claimed he could persuade the emirates to extradite Osama bin Laden. After meeting with President Clinton and Sandy Berger, Ijaz said he could not convince them to extradite bin Laden. Ijaz became a Fox News correspondent and pushed for war in Iraq.

WTC tenants with access to the kind of explosives needed in the demolition of the three WTC buildings did not take into account that NASA uses thermite in its launching of spacecraft and more than 250 strategically placed and precisely timed explosive devices detonate on the space shuttle each time one flies. Lockheed Martin is a major NASA supplier. The thermite is used for space separation and it is done by electronic signals from NASA's space centers around the world.

Crescent's Maurice Sonnenberg was Vice Chairman of L. Paul Bremer's *National Commission on Terrorism* and a member of President Clinton's Foreign Intelligence Advisory Board from 1993 to 1995, with Stephen Friedman of Marsh and McLennan. At the same time, Sonnenberg was a member of the U.S. Commission on Reducing and Protecting Government Secrecy, with Lee Hamilton.

Michael Braham, became CEO of AMSEC in 2003 and was former Senior Vice President at Marsh and McLennon's Crisis Consulting. His boss was L. Paul Bremer. When Marvin Bush joined its board in 2000, AMSEC's annual revenues were $100 million. By 2007, its annual revenues were $500 million.

Barry McDaniel became the COO of Securacom early in 1998. He had worked for the US Government from 1970 to 1987, to ultimately become Deputy Director of Readiness for the United States Army Materiel Command (AMC) located at Fort Belvoir, Virginal; AMC is a primary provider of technology and equipment and explosive ordnance, to US Army personnel. Between 1989 and 1998, he was with BDM International as Vice President of Material Distribution and Management Systems.

After 9/11, Barry McDaniel, who was then CEO of Securacom, was asked whether FBI or other agents had questioned him or others at the company about their security work related to 9/11. He answered "No." The FBI briefly considered investigating Securacom for insider trading related to 9/11 and a SEC referral on suspicious accounts. But as the FBI considered people involved with the *insider trading* had no ties to Muslim terrorists it did nothing. Doing nothing is covering up and the FBI under Louis Freeh was good at that.

Investigations in Washington are many times a cover-up. Congressional committees are convened like a grand jury to go over the evidence and grill witnesses and they can send people to prison for lying to the committees.

Destroying a whistleblower's creditability is standard procedure with the corrupt businessmen, corrupt politicians. There is no doubt that 9/11 was a conspiracy in which thousands of innocent people were murdered just to blame Muslims. Oil and gas are enormously important to the world's financial stability and Jews do the marketing of oil and gas from Muslim countries. Jews have suffered from the Inquisition, Pogroms, and discrimination by Christians but never by Muslims until after the attacks on Muslims by Zionists in Israel.

## Silverstein Properties `

Alan Reiss of the PANYNJ had been working on a three-month transition plan with a team including Silverstein Properties. A few weeks before 9/11, Silverstein Properties asked Reiss to let it operate all systems, from safety to tenant relations. Silverstein had hired Geoffrey (Jeff) Wharton to run the WTC complex for him. Wharton came to Silverstein Properties from Tishman Speyer, one of NYC's biggest office landlords. Wharton was in charge of the buildings when they were destroyed, and remained with Silverstein for one year.

Jeff Wharton was the first person who told Larry Silverstein about 9/11. Silverstein watched 9/11 televised and said to be distressed by the loss of four of his employees. But he was already delving into complex legal strategies to make sure he was promptly compensated.

Jeff Wharton was a friend of Jerome Hauer and through Hauer, Silverstein met and hired an FBI agent named John O'Neill to run security. O'Neill began his FBI career a support employee and worked his way up to assistant special agent and section chief in charge of the counterterrorism division. In O'Neill's 31-year career with the FBI he investigated nearly every terrorist attack ever attributed to Al Qaeda, under close supervision of FBI Director Louis Freeh. O'Neill was part of the investigation of the World Trade Center bombing in 1993 (blamed on blind Sheik Omar Rahman), the Oklahoma City bombing in 1995 (done by CIA agent, Timothy McVeigh, who did the bombing). Dimitri Khalezov says the embassy bombings in Kenya and Tanzania in 1998 and October 12, 2000 were done with mini-nukes. So who was behind, the *suicide boat bombing* of the USS Cole in the South Yemen port?

O'Neill liked to dress like a gangster, fraternize with them, and date several women at the same time lying to all of them. A few weeks before 9/11, he became serious about returning to his Catholic faith and began going to mass every day. He said he felt something was going to happen. On his second day with WTC, O'Neill escaped the building but returned to help others and died in collapse of the building.

In 2007, Larry Silverstein was awarded a $4.55 billion settlement in insurance payouts for the destruction of the WTC, the largest insurance claim ever made.

## Giuliani, Cherkasky, Jack Blum, and BCCI

Kroll's executive VP for Operations, James Bucknam, worked for Rudy Giuliani when Giuliani was in charge of investigating BCCI as SDNY's District Attorney from 1983 to

1989. Giuliani received documents during this time, about secret bank accounts related to terrorist accounts that were in Citibank, Barclays, Credit Lyonnais, and the Japanese company Nomura. Giuliani went to work for White and Case after investigating BCCI yet White and Case was the law firm that defended BCCI.

Giuliani never mentioned terrorism as an issue in his campaign for NYC mayor in 1993. He mentioned the 1993 WTC bombing only once and began looking for a job as police commissioner and would up as Mayor of New York City.

Michael Chertoff is a dual Israeli/American citizenship. His uncle in Israel is a Mossad. Michael was given a top job as Homeland Security Secretary when he should be investigated for his role in 9/11. Can dual Israeli/Americans hold security jobs?

Michael Mukasey, a SDNY US District Judge replaced Alberto Gonzalez as Attorney General in 2006. Mukasey presided over the trial of the 1993 WTC bombers blind Sheik Abdul Rahman who recruited the Mujahedeen for the CIA who trained them to fight against the Soviets in Afghanistan. The trial was so fraudulent Mukasey should have been indicted and sent to prison.

The 1,400 Mujahedeen volunteers fighting in Boania and 2,000 in Chechnya were given the name Al Qeda by the CIA. Osama bin Ladin was called the head of Al Qeda because he paid for air fare of Mujahedeen to Bosnia and Chechnya. Al Qeda does not exist. And there were two Chechens ast bodyguards of bin Laden's widows who were killed by Navy Seals in the *raid to kill Bin Laden.* But died in December 2001 of kidney failure and was buried. 24 Navy Seals in that rid were dead within one year after the *raid* and the 23 year old leader of that raid committed suicide.

On September 4, 2001, President George W. Bush appointed *Robert Mueller as head of the FBI who was in the early 1990s in charge of the Justice Department's criminal division that ran the investigation of BCCI.* Muller was assistant DA of New York County and involved in the BCCI investigations, Michael Cherkasky was assigned the job Jules Kroll worked during the late sixties and early seventies. Michael was also a chief of investigations in the 1993 WTC bombing and became an employee of Kroll when Kroll got the contract to do the WTC security plan (1994 to 1997) for a cool five hundred million dollars. Cherkasky became President of Kroll, then COO, and ultimately the company's CEO, from 2001 to 2004. Kroll's involvement in 9/11 grows.

Jack Blum, former chief counsel to the US Senate's subcommittee's on Drug Trafficking, Law Enforcement, and Foreign Policy, went to New York City and met with the staff at the New York County DA's office. Blum had information about criminal activity *involving BCCI* that was originally about laundering drug money. Oliver North used BCCI to launder dirty drug money. So there was the need to cover-up North's activities that almost ruined the Reagan presidency.

Jack Blum provided information about BCCI and said "the whole Third World was involved. BCCI bought and sold entire governments and maybe United States officials." In July 1991, audits by Price Waterhouse UK said there was massive fraud at BCCI to push the DA to investigate formally. As a result, in July 1991 the New York District Attorney indicted BCCI. Osama bin Laden's BCCI account was wiped out although the FBI has no evidence linking him to 9/11.

The companies and people responsible for revamping the security systems for the WTC buildings and securing New York City on 9/11 shows many have benefited from the investigations of corruption in BCCI in Luxemburg. BCCI was used by the Saudi government, royal family of Kuwait, and Emirate businessmen as BCCI was the only bank in Europe owned and run by Muslims. Oliver North's ambition to make General in the Marine Corps was involved in drug trafficking and used BCCI to launder dirty money. No BCCI officer has gone to prison and neither has Ollie.

PANYNJ occupied a good deal of space in the towers, and some of its twelve commissioners gave false information to NIST for the WTC report and some were linked to BCCI.

Some management representatives at the primary WTC security companies also benefited from the War on Terror such as General James Abrahamson, Terry McAuliffe, Marvin Bush and Larry Silverstein. Although Zionists framed young Saudis as pilots of non-existent planes that hit the North Tower and the Pentagon. The Saudis were tenants of the WTC (and the pilots were supposedly Saudi citizens?).

Four primary contractors involved in installation of the security systems for the WTC had business with Saudis. Electronic Systems Associates' parent, S&H, designed King Saud University, and E.J. Electric worked for Saudi Arabian Airlines. Ensec was owned by a former Saudi arms dealer and Securacom was owned by a member of the Kuwaiti royal family in partnership with Ahmad N. Al Binali & Sons Co., a Saudi Arabian construction company.

The leaders of NY City and NY County played important roles on 9/11 and the ensuing investigations were also involved in the investigators of BCCI and the 1993 WTC bombing and all have benefitted from 9/11. After helping Kroll avoid prosecution for fraud, Cherkasky became CEO of Marsh McLennon replacing L. Paul Bremer who was made the governor of Iraq in charge of rebuilding it.

David Corn and Michael Isikoff of the CIA stated that Saddam Hussein was working on nuclear weapons because aluminum tubing was sent to Iraq.

Valerie Palme was a top ranking CIA officer and the CIA ssent her husband, Joe Wilson, to Niger, Africa to see if Saddam had bought any yellowcake (uranium ore). Wilson reported that Saddam had not bought any yellowcake. President Bush's White House put pressure on Wilson to recant and support Corn and Isikoff's claims but Wilson refused. But when Valerie's

113

appeared in Robert Novak's column who as a top CIA officer her cover was blown endangering her life. So Carl Rove in the Bush White House assumed was blamed. But was he just the scapegoat for the President or Vice President Cheney?

Hollywood produced *Medusa's Head* in 1978 that had a Boeing airliner fly into a tall Skyscraper. And it happens on 9/11.

CIA relies on the ignorance and trust of politicians and lower level; scientists to convince them that the aluminum that the tubing imported by Saddam Hussein was to make centrifuges. No, it was for Iraq's missile program.

In 1994, special agent Robert Wright said the FBI was *obstructing investigations* into the financial dealings of Al Qaeda. After the break-up of the Soviet Union December 25, 1990, Yugoslavia began breaking up and Serbs began killing Bosnian Muslims. The British and French blocked efforts to allow Bosnia to buy weapons for its defense and Strobe Talbot (Bill Clinton's roommate at Oxford) said supporting Boris Yeltsin as the first president of Russia was more important.

The Russian military was planning to remake the Soviet Union and assisted in the massacres of Bosnian Muslims. Osama bin Laden paid the way for Mujahedeen volunteers to fly from Afghanistan to Bosnia and by mid-1964, they numbered 1,400 and were rapidly clearing Serbs out of Bosnia using the psychological warfare tactics the CIA taught them. There is no organization called *Al Qaeda*. It is a CIA invention as those Mujahedeen referred to Afghanistan as Al Qaeda, *the Base* and no world-wide *Al Qaeda* financial system. The only financing came from billionaire Osama bin Laden.

Saddam Hussein came to Bosnia and offered WMDs which Bosnian leader, Izetbegovic. He lied to say Bosnians had created them. Saddam then became a target for elimination.

The CIA had recruited the Mujahedeen through Sheik Abdul Rahman. But as Osama bin Laden acted on his own without consulting the CIA and Washington, he became the enemy. That is why the Emirates, Kuwait, and Saudi Arabia are follow orders from Washington and support Israel.

The Mujahedeen came to Chechnya to help them fight Russia. Boris Yeltsin ordered atrocities on Chechen villages, but 20 Russian generals refused to massacre villages even though Yeltsin and General Grachev threatened to execute them.

As Russian losses grew, the name Al Qaeda was used more internationally. Osama bin Laden had visited Chechnya and paid the transportation for hundreds of Mujahedeen to come to Russia. Two Chechens became his bodyguards. After he died in 2007, they stayed to protect his wives. They were killed by the Navy Seals who then dug up bin Laden's body and buried it at sea to hide the fact he was already dead.

The CIA put a grainy video of Osama bin laden laughing with a Mujahidin leader on television with the translation being, "and the walls will come tumbling down and thousands will be killed," as evidence Osama bin Laden was behind 9/11. It was previously shown on TV with a different translation. So more fake evidence. Then the fake raid on *Osama bin Laden's hideout in Pakistan was:* 1. to raise Mr. Obama's popularity in the polls, 2. Point suspicion at Pakistan to get more cooperation, 3. Make Mr. Obama submissive with the threat of revealing it was all fake and 4. give the military more power in decisions on diplomatic and military policies. Just one year after Navy Seals were said to have *killed* bin Laden, 22 were dead and the leader of the raid *committed suicide.*

Benazir Bhutto of Pakistan was killed by bombs thrown by assassins on December 27, 2007. One assumption was that Al Qaeda assassinated her. Bhutto knew Osama bin Laden was dead and so did Pakistan President Pervez Musharaf who was also suspected to have had her killed. Losses from the destruction to Pakistan's economy from the riots that followed Benazir's murder was 8% of Pakistan's total GDP that year.

# How Israel Creates Terrorists

After William Rodriguez, the janitor of the North Tower was interviewed by BBC TV in London the British public pushed for an Inquiry of why Britain went to war in Iraq. It was broadcast on BBC TV with ex-Prime Ministers Tony Blair and Gordon Brown grilled on the war, Their response, "according to the best intelligence we had, Saddam Hussein was an immediate threat to the West." But Tony Blair *suppressed* a British Intelligence Report that the Mossad unit of six Egyptian Jews, set up Atta-Mamoun Darkanzali and his friends to be *suicide pilots* on 9/11. Tony Blair knew 9/11 was an Israeli False Flag Operation and if Tony knew that meant George W. Bush knew and George ordered the torture used on Sheik Khaled Mohammed (page 42) to make him confess he was the Mastermind of 9/11. Another detainee died from torture.

Germany under Hitler permitted the International Red Cross to inspect all of Germany's concentration camps and prisons from 1933 to June 1945 and talk to any prisoner. But the United States has refused to allow the International Red Cross to inspect the prison facilities at Guantanamo and access to the prisoners. Count Folke Bernadotte of Sweden was in charge of inspections of German concentration camps and prisons and was assassinated September 17, 1948 by the Israeli Lehi terrorists, to which future PM Yitzak Shamir belonged.

Setting up Mohammed Atta and friends was like a page from the Assassins where an assassin is coached to make a friend of the intended victim. The six Mossad are divided into two thee-man units. One unit headed to Schiphol International Airport at Amsterdam where the Mossad offices are located in El Al airline offices. The other group headed to Hamburg to meet Mohammed Atta and his friends to tell them that Osama bin Laden sent them and wants to fly to America.

But it is Ephraim Halevy, Chief of the Mossad, who pays for their tickets to America. The two teams call Mohammed and his friends an Al Qaeda cell. If they were in any way connected to Al Qaeda they could have thanked Osama bin Laden personally. They fly to Logan Airport, then fly to Miami and then to Hollywood (Florida), Vero Beach, Delray Beach, and West Palm Beach. There they came under surveillance by the DEA that suspected them of drug dealing. In mid-August, the Mossad team is in Boston reported to Tel Aviv that the Florida team had *completed pilot training*.

The two Mossad teams sent coded reports regularly on the progress of their operation *via the Israeli Embassy in Washington, D.C. which sent them on to Tel Aviv* and all avoided any mention of the World Trade Center. On September 10, 2001, leading news media people were called to the Israeli Embassy in Washington and instructed to pin 9/11 on Arab hijackers, Osama bin Laden, Al Qaeda, and the Taliban.

In August 2001, the first Mossad team flies with Mohammed Atta and his Hamburg "Al Qaeda" cell to Logan International Airport. Logan's security is Huntleigh USA, owned by International Consultants on Targeted Security (ICTS) which is an Israeli firm whose owners are friends of Israeli President Benjamin Netanyahu.

At Logan. Mohammed Atta supposedly went through immigration easily even though he was deported from the US to stand trial in Israel, imprisoned and then released. Logan Airport is where they are supposed to board American and United Airlines flights on September 11.

Mossad Chief Ephraim Halevy tells George Tenet, Director of the CIA that the Mossad *heard* about a threat by Arab terrorists to bomb a nuclear plant somewhere on the East Coast. French and Egyptian intelligence also sent information of a threat by terrorists. George Tenet, head of the CIA sends an unspecified warning to the new FBI director, Robert Mueller a week before

September 11, 2001. A routine FBI bulletin is sent to field agents to look for any sign of a terrorists attack but no high alert. When George Tenet hears the news of 9/11 he says, *This has Osama bin Laden's fingerprints all over it.*

The *Le Figaro* newspaper of Paris reported two months before September 11, Osama bin Laden was visited by two CIA agents in a Dubai hospital where he was undergoing kidney dialysis. had made him a hero for bringing his machinery from Saudi Arabia to make fortresses in mountains of Afghanistan for the Mujahedeen the CIA trained to free Afghanistan from Godless atheist Russians. At stake was the huge gas reserves of Afghanistan the Russians needed to improve their economy now were be piped to the Indian Ocean via Pakistan and sold to China and Japan. The CIA training the mujahedeen needed bin Laden as he was only one who could speak English.

The CIA then asked British Intelligence for permission to use the Muslim Brotherhood (Ikhwan), which the British have always controlled, and the British gave them the blind Egyptian cleric, Sheik Omar Abdel Rahman. He recruited a Muslim Foreign Legion that the CIA trained, armed, and financed in Afghanistan (called the Mujahedeen) to fight the Russians who wanted to annex the nine northern provinces of Afghanistan that held huge reserves of natural gas that would have saved the Soviet Union what was going bankrupt.

The CIA then taught them psychological warfare that drove the Russians out of Afghanistan. Filipino Muslims returning to the Philippines used these tactics and Filipinos resent what the CIA did. When Osama bin Laden paid the air fare for Mujahedeen to travel to Bosnia and Chechnya and fight the Serbs and the Russians, the CIA under Bill Clinton called then the *Al Qaeda* and the CIA and FBI framed Sheik Omar Abdel Rahman with blowing up the WTC North Tower in 1993 which was done by the CIA and FBI under Louis Freeh.

The Bosnian War began just after the break-up of the Soviet Union. The new Russian Federation was saddled with debt and almost bankrupt. The Russian military decided to re-create the Soviet Union and reoccupy the Ukraine, Poland, and the Balkans and just like the British used Yugoslavia to menace Hitler's right flank when he was poised to invade Russia, May 1, 1941, Russia needed Yugoslavia to menace NATO's right flank if NATO opposed Russia,

Yugoslavia was breaking up in 1991 because of economic problems and Slobodan Milosevic was financed by the Russian military. Chancellor Helmut Kohl forked over 21 billion Deutchmarks for the Russian investments in East Germany and 59 billion to build housing for Russian soldiers to unify Germany. So Serb snipers in Sarajevo were paid 500 DMarks for every man, woman, and child they killed and not dollars.

Russian Minister of Defense, General Grachev, needed Yugoslavia to menace NATO's right flank when the Russian Army would re-enter Ukraine, Poland Czechoslovakia, and the Balkans to remake the Soviet Union. Yeltsin was the president of Russia but without Grachev, he owed Grachev who was planning remake the Soviet Union with the leadership in the hands of the military. He was going to subvert Yeltsin but a top Russian intelligence agent, Vladimir Putin unmasked Grachev and became Yeltsin;s successor. A strong Yugoslav Army was needed to threaten NATO's right flank to block Grachev's move into Ukraine, Poland, and the Balkans, but it was breaking up from economic problems. So a campaign of Serb atrocities were planned to destroy Bosnia that was accepted into the United Nations.

Croatian Serbs declared independence first but its Army was no match for the Yugoslav Army and thousands were massacred. The UN created a ceasefire that held because the Serb invasion had to pass through a narrow corridor between Hungary and Bosnia to get at Croatia. Thus Bosnia with its longer border with Croatia brutalized by a campaign of rape, murder, arson, torture, and massacres.  Chancellor Helmut Kohl gave the Russians 8o billion Deutschmarks to the Russians to get out of East Germany.  Bosnian Serbs in Sarajevo were given sniper rifles and paid 500 Deutschmarks for every man, woman, and child killed.  Those Deutschmarks were part of the money Kohl paid the Russians. .

Secretary of State James Baker put an arms embargo on Yugoslavia, Bosnia, and Croatia through the UN Security Council which did not affect the Serbs and Milosevic discharged 80,000 Bosnian Serbs to make a Bosnian Serb army with tanks, trucks, jet planes, helicopters, fuel and all supplies and called it a civil war. It was a war of aggression by the Serbs which had thousands of Russians and Greeks flocking to join the Bosnian Serb Army and fight Bosnia a member nation of the United Nsations and the UN Security Council did nothing.

The Russians sent their all their military equipment from East Germany to Serbia.  The plan for the Russians would be to fly in 20,000 troops to take those weapons and push into southern Poland with Russian troops pushing into Poland from the Ukraine, Baltic states, and north to recreate the Soviet Union.

Greece is a member of NATO but 8,000 Greeks volunteered to fight for the Bosnian Serb Army and 6,000 Russian volunteers were supplied by their own countries. The Mujahedeen from Afghanistan that came to Bosnia and given the name *Al Qaeda by* the CIA and numbered only 1,400.  They were rapidly clearing Serbs from Bosnian land using tactics the CIA taught them.  Just mentioning the word *Bosnia* to Boris Yeltsin he'd scream he would enter the Bosnian war on the side of Serbia, and Strobe Talbot did everything to mollify him which was to destroy Bosnia.

After Colonel Ratko Mladic massacred three thousand Christians at Vukovar, he was promoted to General of the new Bosnian Serb Army,  Slobodan Milosevic of Serbia then released

80,000 Bosnian Serbs to give General Mladic his army. Milosevic also gave Mladic 512 tanks, 506 armored personnel carriers, ten high performance military jets, more than 250 mortars, howitzers, and other artillery, 18 transport helicopters, and 8,700 tons of fuel while Bosnia has one rifle for every three men. James Baker and George H. W. Bush helped the Russians and Clinton's Under Secretary of State, Strobe Talbot wanted to promote Boris Yeltsin as the first president of Russia, so news about Bosnia was kept from the American public.

Although American Jews sympathized with Bosnian Muslims, Israel was on the side of Serbia. No Jews served in the Bosnian Army, but Jews served in the Bosnian Serb Army and one did the Srebrenica murder of 8,700 men and boy.

Returning to how Israel creates terrorists, Ramzi bin ash-Shibah, born May 1, 1972, is a Yemeni and detained in the Guantanamo Naval Base in Cuba. He is accused of being a key facilitator for the September 11, 2001 attack.

In the mid-1990's Ramzi went to Hamburg, Germany where he became friends with Mohammed Atta, Ziad Jarrah, and Marwan al-Shehih which the Mossad called the Al Qaeda Hamburg cell. Ramzi was identified by the US as the 20th hijacker and had been denied entry to the United States.

Captured in Karachi, Pakistan on September 11, 2002, far from any battlefield, Ramzi was accused of wiring money and passing messages from Al Qaeda figures and charged with being an enemy combatant. He was tortured with sleep deprivation, stripped naked and exposed to the cold and George W. Bush's White House legal counsel, Alberto R. Gonzales, said this was not torture and President Bush and Vice President Cheney agreed.

After being *waterboarded* 183 or 187 times, Sheikh Khaled Mohammed confessed he was the Mastermind behind 9/11. *Waterboarding* is having a cloth placed over the face with a constant stream of water pouring down and hands are tied. One detainee at Guantanamo died undergoing this torture. Dick

Cheney who shot his 87 year old friend *while hunting* supports torture even if a person were innocent

Judge Susan Crawford, a lifelong Republican, said she sympathized with the interrogators trying to extract information from the detainees about further plans to attack America. Judge Crawford said *there is a fine line that must not be crossed in interrogation.* Crawford also said, The only trouble we have with the prisoners is that they are gaining weight from the good food they are getting." Well, no food passed over their tongue. All were held down and a tube roughly forced into their rectum and force-fed. Join the Marines and learn a vocation. They will need mental help in returning to civilian life.

Wild Bill Donovan was the most highly decorated man to serve in the US military and sent to London on a special mission. He returned to recommend centralization of US Intelligence. So on July 11, 1942, Donovan became *Coordinator of Information (COI)* to centralize intelligence as State Department, FBI, Navy, and Army had jealously competed instead of collaborating.

In November, 1941, he initiated a research program for *truth drugs* to use in the interrogation of spies and agents. S*copolamine*, cannabis, mescaline, tetra-hydro-cabbabinol acetate and opium were already in use in psycho-analysis by psychologists and Sodium Pentothal (*Truth Serum) was* used by both Germany and Britain. When the British evacuated their soldiers from the beaches of Dunkirk, 100,000 were shell shocked and useless. They were put to sleep with Pentothal, put in beds in huge underground bunkers and intravenously fed for two weeks. When they were awakened their they were cured and returned to active duty. Some died in their sleep.

The CIA Mkultra mind control experimented with drugs that were to reduce people into zombies that would do anything the manipulator wanted them to do. Head of CIA, Richard Helms destroyed Mkultra records in 1973.

A defector from Russia's biological weapons Department 12, of the KSB, presently part of Russia's SVR service, claimed that a truth serum named *SP 117* has no taste, no smell, and no color, no immediate side effects, and fell asleep with no recollection.

Alberto Gonzales, White House legal counsel, said, "The international laws outlawing torture are antiquated and unrelated to today's world." He authorized *enhanced interrogation* (rough forced feeding food through the anus, water-boarding, and naked in a cold room for 24 hours or more. Vice President Cheney agreed and President Bush approved it and it became a White House directive.

Gonzales was then made Attorney General of the United States and made a professor of law at Harvard University.

CIA Director George Tenet established CIA's own torture at Guantanamo. The author knows of the persecution of Muslims by London police and British intelligence, the United States, Israel, and Russia in an effort to induce Islamic extremism such as the Mossad-MI6 *Hornet's Nest.*

In his final address to the nation, President Dwight Eisenhower told the nation in his farewell address to America that greatest danger to America is the military industrial complex. Eisenhower was a general and knew the military industrial complex loves money. In WW II a jeep cost $500. Jeeps were replaced by Hummers that cost $50,000. The cost of an atom bomb rose from $20 million to 200 million. The floors where the plane struck the North Tower were offices occupied by military weapons suppliers.

From *Snopes.com* comes a story about Mohammed Mahmoud Atta, also known as Mahmoud Atta Amadi, who was named as the hijacking pilot of Flight 11. Atta, a 33 year old Jordanian native on April 12, 1986 opened fire with an Uzi, on a bus in Israeli Occupied West Bank, killing the driver and seriously wounding three passengers.

Atta was arrested in Venezuela a year later and as a naturalized US citizen, extradited to the US where he was arrested by the FBI and held for in prison for three years claiming his crime was political. In 1990, he was extradited to Israel after the US Courts upheld Israel's request for extradition as the bus was civilian and not a military target.

In October 1991, an Israeli Court found Atta guilty and sentenced him to life imprisonment. George Schultz who was Secretary of State under George H.W. Bush, wrote that President Clinton and Secretary of State, Warren Christopher, requested Israel to release political prisoners and insisted that Atta be released with them. Warren Christopher resigned in 1985 and was replaced by Madeleine Albright.

After September 11, 2001, the investigators of 9/11 concluded that Mohammed Atta was the Mastermind behind the hijacking of four commercial airliners. Atta had entered the US and spent over a year learning how to fly an airplane at more than one flying school and still could not fly a Cessna according to one flying instructor who said that he himself could not fly a Being 767 into the South Tower even if he took over the controls after it was airborne.

The *Australian* in 2001, asked how could Atta who was a prime suspect in a 1986 terrorist attack on an Israeli bus get into the US? The answer that Logan Airport security was controlled by an Israeli company and the Mossad.

The *San Francisco Chronicle*, 2001, referred to the most chilling articles in the *Los Angeles Times* never explained how Mohammed Atta traveled in and out of the United States on expired visas and was on the government's *Watch List* linking him to a bus bombing attack in Israel.

According to The *Jerusalem Post*, Atta "was freed after the Supreme Court ruled there were faults in the extradition process." Newspaper accounts as late as 1993 said Atta was "serving a life sentence in an Israeli prison."

Then in mid-2002 the Jerusalem Post disputed the claim that Ollie North warned everyone about Osama bin Laden with a statement that said "No, Ollie warned about Abu Nidal not Osama bin Laden." Abu Nidal was killed in a shoot-out at his Bagdad Apartment on 16 August 2002 and died by suicide or at the hands of Saddam Hussein's secret police at age 65. Abu Nidal was noted for terrorist acts in 20 countries killing over 300 people but none in Israel. Many suspect Abu Nidal of being an Israeli operative as he killed and tortured more of his volunteers than people in his terrorist attacks. His bombing of a Berlin café that killed some American soldiers was aided by an East Berlin Vopo who worked for Israel. Israel blamed the bombing on Muamar Gadaffi, the leader of Libya. So President Reagan sent Tomahawk missiles into Gadaffi's home in Tripoli that killed a daughter and went before the United Nations to proclaim "America will strikes at its enemies anytime and any where," but the information was false as it turned out.

Gadaffi was killed by a French assassin on orders of Sarkozy after Gaddafi was captured by the rebels. Gaddafi had given Sarkozy one million dollars for his re-election campaign and Sarkozy wanted to hide that from the news media.

Gadaffi's sons were killed and a surviving daughter, Ayesha, who is a human rights lawyer, has been given sanctuary by Oman's Sultan. The new Libyan Rebel Government used INTERPOL to issue a warrant for her arrest. Algeria's leader refused to give her sanctuary or tallow her to stay in Algeria even though she over seven months pregnant.

Sources: The *Boston Globe*: Burger, Timothy J. "9 Hijackers Were Legally in Country." 19 September 2001, p. A32

*Confronting the Terror: Plot May Have Been in Making for 5 Years.*" by Cullen, Kevin

The *Jerusalem Post*. Kurtz, Howard. "U.S. Citizen Held for Raid on Israeli Bus." 1 September 1991

The *Washington Post*  Lardner Jr., George.  "U.S. Extradites Alleged Terrorist to Israel." 7 May 1987  p. A43.

The *Jerusalem Post*,  "Justice Ministry: WTC Bomber Was Never Held in Israel." 7 November 2001

*The Australian.*  "CIA Incompetence Is Criminal."  18 September 2001  (p. 3).

The *London Independent.*  "Terror in America: The Face of Hate." 14 September 2001  (p. 1).

The *Jerusalem Post.*  "Terrorist Gets Life for Bus Attack." 2 October 1991.

*Wikepedia*: The free internet encyclopedia,

Mohammed Atta, the alleged pilot of Flight 11, did not go through security at Logan on September 11, 2001.  Could have Flight 11 been an Evergreen Boeing 767 painted to look like an American Airlines passenger plane.  Phillip Marshall, who flew for the CIA, revealed that Evergreen painted planes to look like commercial airliners for the CIA, but Marshall was killed in California along with his two teenage children. The local sheriff's office called it a double murder and a suicide and the sheriff had the crime scene cleaned by professional cleaners before Crime Scene Investigators went over the house.  All Phillip Marshall's files and computers were removed by the FBI.

Rebecca Roth, a retired flight attendant, says that Amy Sweeny and Betty Ong the two Flight 77 flight attendants, were not doing what they were supposed to do at time of the reported calls they made to American Airlines offices at North Logan Airport, Dallas, and North Carolina.  Rebecca is positive the two were not on any plane as passengers would be screaming when the plane dove down from 10,000 feet to hit the tower.

On the morning of September 11, 2001, Jennifer Oberstein was exiting Bowling Green subway exit and as she looked up at the North Tower, orange flames shot out from upper floors.  Taking out her cell phone she immediately called her friend, Katie Couric. Katie was broadcasting on NBC's "Today" show with co-host Matt Hauer and they had just been informed by the news room a plane hit the North Tower. But when Katie asked Jennifer, "What kind of a plane was it that hit the tower?" Jennifer responded, "Plane? What plane? There was no plane!"

The NIST report does not mention anything about the phone call Jennifer Oberstein made to Katie Couric.

According to the Eglin Air Base laboratory the steel used in construction of both towers was inferior steel which is against building for hi-rises buildings. Larry Silverstein built the whole WTC in 1972. If Larry Silverstein used inferior steel in construction of the WTC he was guilty of manslaughter but Attorney General Ashcroft did nothing.

## Francesco Cossiga: 9/11 a CIA/Mossad Operation

Former Italian President, Francesco Cossiga, revealed the existence of Operation Gladio and in an interview with Italy's oldest and most widely read newspaper, *Corrierre della Sera* that 9/11 was a Black Ops done by the CIA and the Mossad. He said that it was common knowledge among the top intelligence agencies around the world that Israel did 9/11 with the CIA.

Six months after 9/11, both Russia and China knew 9/11 was a Black Ops. Most Muslims did not believe 9/11 was done by Muslims but their governments supported the official report as true as did Australia, India, Canada, all Latin America. Some Muslims who did believe in the official report reacted in a way to justify 9/11 as a result of injustices such as the UN and Western nations support for Serbs who massacred, raped, and murdered unarmed Bosnian Muslims. Germany forbids any introduction of anything about 9/11 to be introduced in German Court. How can Arabs and Muslims in Germany when German Courts forbid and accused of being involved in 9/11 can defend themselves? Evidence from professional engineers, pilots, and architects that would prove Muslims are innocent is barred.

Francesco Cossiga was first elected president of the Italian Senate in 1983. He then won a general election in 1985 and serves as president until 1992 when his participation in Operation Gladio forced his resignation. Operation Gladio was a covert intelligence network operating under NATO to carry out False Flag-Black Ops terrorism attacks to be blamed on European political groups in March 2001. Gladio agent, Vincenzo Vinciguerra gave an affidavit where he said, "You had to attack civilians, the people, women, children, innocent people, unknown people far removed from any political game was to push the public to ask the state for greater security."

Writer-researcher Webster Tarpley quoted Cossiga as saying, "The Mastermind of 9/11 had to have a sophisticated mind, and the means not only to recruit fanatic kamikazes but highly specialized personnel. I add one more thing. It could not be accomplished without the infiltration of the radar and flight security personnel." Francesco Cossiga knows from his experience in Operation Gladio that 9/11 was far beyond bin Ladin's capabilities. Only Israel could have done it.

## Many Accused 9/11 "HIJACKERS" Are Alive

While the news media has focused on the suicide hijackers being Muslims, many of them identified by the FBI are still alive and innocent as reported by British BBC/

Abdul Aziz Al-Omari (Flight 11) (Trained Pilot). The FBI cobbled the names of two men together to create a terrorist but both are Alive! The first has the same name, the same birth date but has no idea how to fly. The second has the same name, Abdul

Rahman Al-Omari, but a different birth date. The person pictured below by the FBI is a pilot for Saudi Arabian Airlines.

v Omari Number 1 is the real Abdul Zaiss Al Omari whose passport was stolen in 1995 while studying electrical engineering at the University of Denver and notified the Denver police of the theft. He said, "I couldn't believe it when the FBI put me on their list. They gave my name, my date of birth, but I am not a suicide bomber. I am here. I am alive. I have no idea how to fly a plane. The New York Times wrote that Al-Omari has been found in Saudi Arabia and is cleared in the case." Saudi Embassy officials in Washington say, "A Saudi electrical engineer named Abdul Aziz Al-Omari had his passport and other papers stolen in 1996 in

Denver when he was a student and reported the theft to police there at the time."

Abdul Aziz Al-Omari (Flight 11) (Trained Pilot). The FBI cobbled the names of two men together to create a terrorist but both are Alive!  The first has the same name, the same birth date but has no idea how to fly. The second has the same name, Abdul Rahman Al-Omari, but a different birth date.  The person pictured below by the FBI is a pilot for Saudi Arabian Airlines.

Waleed Al-Shehri (Flight 11) is a professional Pilot and is the sixth person on the FBI's list, He is a Saudi national living in Casablanca and works with Royal Air Moroc.  He had lived in Dayton Beach, Florida where he took flight training at Embry-Riddle Aeronautical University.  Al Shehri spoke to the U.S Embassy in Morocco and told them that he was not a member of the suicide attack nor a terrorist.  "He was reported to have been in Hollywood, Florida, for a month earlier this year but his father, Ahmed, said that Waleed was alive and well and living in Morocco."  He has never lost his passport.  The FBI said his passport was "stolen" without checking.

Abdel Aziz Al-Omari, Omari Number 2 is a pilot with Saudi Airlines. He went to the US Consulate in Jeddah to demand why he was being reported as a dead hijacker in the American media. He lives with his wife and four children in Jeddah.

Adnan Bukhari "Adnan Bukhari is still in Florida."

Amer Kamfar  was named by the FBI as a suicide bomber is a pilot and alive in Arabia."

Amer Kamfar  was named by the FBI as a suicide bomber is a pilot and alive in Arabia."

Adnan    Bukhari    "Adnan    Bukhari    is    still    in    Florida."

Saïd Hussein Gharamallah Al-Ghamdi, (flight 93) is a professional pilot and alive and at his job for Tunis Air.  He said, "I was completely shocked. For the past 10 months I have been based in Tunis with 22 other pilots learning to fly an Airbus 320.  The FBI has not provided any evidence of my involvement in the attacks." US officials have said he was linked to *Osama bin Laden's al-Qaeda network."*

Despite world news media that they are alive, Robert Mueller's FBI has not made any corrections for wrong names, wrong photos, and wrong identities as the FBI alleges the hijackers used false identities.  Louis Freeh was appointed head of the FBI in 24 hours after Clinton fired William Sessions as head of the FBI.  Louis Freeh resigned on May 4, 2001 and it took four months to replace him with Robert Mueller on September 4. 2001.  The Mossad supplied the list of hijackers to the FBI with all the details of how Mohammed Atta and his friends flew from Germany to the US to learn how to fly Boeing 767's.  That's how the FBI knew all about Mohammed Atta and his suicidal hijackers.

To begin with, the WTC attacks were described as *brilliantly planned and executed by professionals.*  The FBI assumed the hijackers used their real names.  How could brilliant professionals, leave a car at an airport found within hours of the attack with a flight training manuals in Arabic?  Were they still

131

reading the 'manuals' on how to fly?" As they weren't planning to come back it they should have taken taxis to the airports.

One of the hijackers carries his passport on a domestic flight which he will use after his suicide bombing to flee the US after committing suicide? Atta's passport survives the fireball, flutters to the ground a few blocks away in good condition. Atta packs a suitcase with a book on "how to be a terrorist in the US" document he forgot to take. He drove from Florida to Portland Maine to fly to Boston. Why by-pass Boston to fly to Boston to catch his Los Angeles flight (Flight 11) that originated at Logan Airport in Boston? Atta would have to go through another security check and screening of his baggage.

*Within hours after 9/11*, authorities found evidence connecting Arab Muslims to 9/11. The *Boston Globe* reported that Authorities were tipped off about a suspicious white car at Boston's Logan Airport where Arabic-language flight training manual was found inside the car and in few days, the FBI identified all 19 hijackers.

Seven of the hijackers named came forward to say they were alive and protested that they felt libeled by being names as involved in 9/11. David Harrison of Britain's *Telegraph* wrote about four innocent men told him how their identities had been stolen, "The men were all from Saudi Arabia and spoke of their shock at being named by the FBI as suicide terrorists." None of them were in the United States on September 11 and all are alive in Saudi Arabia and outraged to be called terrorists. One had never even been to America and another, a Saudia Airlines pilot, was in Tunisia. Saudia Airlines considered taking legal action against the FBI for damaging its reputation and its pilots.

On September 21, 2001, CNN reported FBI Director Robert Mueller admitted that the names of innocent men who are alive have not been removed from the FBI list,.

Metal detectors at Logan Airport's departure gates would have detected a knife or even a small bullet while the X-ray machine screening would do the same let alone box cutters. /There are no video surveillance cameras at any of Logan's checkpoints and none of the security people recalled seeing any of the Flight 11 hijackers, or reported anything suspicious yet they drove all the way from southern Florida to Portland, Maine (passing Logan Airport) to take a flight to Logan where they were to change to Flight 11 to California,

How could Arab students take a simulator course and then fly jumbo jets with the skill and precision of top pilots? How could they have overtaken pilots in the cockpits who were bigger and stronger? If a plane is reported as hijacked the facility at Pike's Peak in Colorado Springs has the capability to take over the controls of 4,000 planes flying over the United States at any time and land them by remote.

American Airlines flight attendant Betty Ong on Flight 11 calls American Airlines Reservation Office in Cary, North Carolina to report that two of the five flight attendants were stabbed. She tells them that two men broke into the cockpit of the plane. Nydia Gonzalez is on the other end of the phone picks up a phone on a separate phone line, to alert the SOC, in Fort Worth, Texas as it coordinates minute-by-minute operation of American Airlines.

Craig Marquis is the manager on duty in Ft. Worth, and Nydia tells him what Betty Ong has told her. Amy Sweeney, the other flight attendant uses another phone to make three calls to the American Airlines flight services office at Logan International Airport in Boston to describe the events on the plane. The first two calls, made at 8:25 a.m. and 8:29 a.m. Amy was disconnected after less than two minutes but her third call, at 8:32 a.m., stays connected 8:45.

At 8:25 a.m., Craig Marquis, the SOC manager called Peggy Houck, a flight dispatcher at the SOC, and asked her to try and contact the pilot of Flight 11. Marquis told Houck that number three flight attendant, Betty Ong, had contacted the airline's reservations office in Gary to report a passenger on board stabbed two flight attendants and added that Ong had been "trying to get hold of the cockpit crew and couldn't get through then and the cockpit cabin door was closed." Marquis added, "Don't spread this around. This is between you and me right now, okay?" Houck answered, "Okay."

When Amy Sweeney made her first call to American Airlines flight services at Logan after 8:25, the employee called the SOC in Ft. Worth and talked to Ray Howland, senior manager at the SOC and told Howland that the flight services office got a call from a flight attendant on Flight 11that the plane might have been hijacked. Howland replied, "We don't want this getting out. We're aware of the situation. We're dealing with it right now. So let us deal with it. We don't want anything getting out right now." The employee replied, "Nothing said. Okay."

Craig Marquis told the manager of SOC policies and procedures, Mike Mulcahy, that *number three flight attendant* on Flight 11 had called and reported number one and number five flight attendant by two male passengers and broke into the cockpit. She said the plane was flying erratically. Mulcahy told Marquis that he wanted all information on Flight 11 brought to him and added: "I don't want this spread all over this office right now okay?"

There were five flight attendants and after two were stabbed, Betty Ong and Amy Sweeny get on the plane's phones to tell about the stabbings and hijackers forcing their way into the cockpit to take over the controls. Betty and Amy talk continuously on the phones to the American Airlines supervisor at Logan until the phone goes dead at 8:45 a.m. when Flight 11 hits the South Tower.  Betty Ong and Amy Sweeny used the words, *Middle-Eastern* looking.  Israelis are Middle Eastern and some speak Arabic. Ong and Sweeny alerted senior American Airlines personnel to an emergency and supervising officers ordered everyone in the office to maintain silence about the hijacking.  A veteran flight attendant for United Airlines said, "This was disgusting.  Their very first response was to cover-up when they should have been broadcasting this information all over the place."

Where are Betty and Amy today?  They certainly weren't on Flight 11 as no plane hit the North Tower.  Before the smoke comes pouring out of the North Tower, a girl is seen waving from that huge hole.  How did she survive?

The three videos that came on television immediately on 9/11 are said to be inserted into the video.  The plane flying past the smoking North Tower and rams into the South Tower is completely swallowed up but no plane parts fall to the ground. Aluminum planes cannot cut through the steel beams all the way to the other side of the building.

**When employees at Logan Airport learned of the hijacking they sent word around on their own to others in Logan Airport to document what they were doing at the time Flight 11 was loading passengers and control tower was clearing the take off.  The Israeli-American CEO at Logan Airport ordered everyone to destroy their documents.**

Tony Szamboti was a mechanical engineer and working in the aerospace industry and was knowledgeable mechanics and stress analysis.  After watching videos of the twin towers coming down, he said it was difficult to correlate the amount of energy needed to see the collapse of the twin towers in a free fall.  In 2006, after he heard the presentation of Physics Professor Steven

Jones showing video where molten metal was spilling out of the corners of the towers. Tony said an autonomous investigation of 9/11 was needed.

## Photos of the Horror of 9/11 and Phone Calls

Marty Lederhandler was an Associated Press Photographer and in 1937 he covered the burning of the giant Hindenburg Zeppelin. In 1944, he waded ashore at Utah Beach on D-Day with two carrier pigeons to carry his undeveloped film back to England. On September 11, 2001, he grabbed his camera and went to the GE Building in Rockefeller Plaza and took the elevator to the 65th floor, the Rainbow Room and went to the big picture window to take pictures of the twin towers. When the order to evacuate the building he said, "The only other story that compares to this is D-Day." Marty Lederhandler died in 2009.

Richard Drew is in the business of photo-ing fashion shows for 35 years. He was on his second week in September was Fall Fashion Week. As a 21-year-old photographer for the Pasadena Independent-*Star News*, he was at the Ambassador Hotel in Los Angeles on June 5, 1968, when Robert Kennedy was assassinated.

On September 11, AP photo editor Barbara Woike called and told him a plane had hit the North Tower of the WTC. Drew rushed to the subway and took the No. 2 train to Chambers Street. Emerging from the subway, he saw smoke billowing from both towers and took a position near a line of ambulances to wait for casualties when a paramedic shouted, "Look! There's people coming out of the World Trade Center!" She wasn't pointing down the street but pointing up. Bodies from the towers were tumbling out fast.  After developing his film, one photo stood out: A man in black pants and a white jacket, one leg bent as he plummeted headfirst.  It would become known simply as "The Falling Man."  Of all the photos he took that day, it is the least published.  Drew thinks "people react to it; because they can relate to that it might be them."

,  ,  ,

Photographer Doug Mills was covering President George W. Bush's was to visit with kids at Emma E. Booker Elementary School.  On the way to the school, Mills heard snatches of the deputy press secretary's cell phone conversation that a plane had hit a New York building.  At the school, Mills and the other journalists were herded to the back of the classroom and Mills began video-taping the president and the children.  Five minutes the classroom door opened, and White House Chief of Staff, Andy Card, stepped inside. Mills made eye contact with Card and mouthed... "What's going on?"  Card held up two fingers.

"We had NO idea what that meant," says Mills. "So, like, Two minutes, we're leaving? ... or He'll talk to me in two minutes ..."

Card waited a moment before he walked to the front of the room, leaned over and whispered into Bush's right ear.  The president's face went blank.  Then everyone was told to go to the motorcade.  As the classroom event was not televised live,  Mills edited the tape and sent it to the AP photo desk.  AP responded by asking what Card did tell Bush.  Doug Mills replied only that it was about planes hitting the twin towers.

Bob Daugherty the AP Washington photo editor then told him, "Great job, kid."  Only after they boarded Air Force One and began watching CNN that the full impact came into focus.  "If the attacks had happened while we were at the White House we would not been present when Andy Card would have entered the Oval Office to tell the president."

Later, Mills asked Card what had he whispered into Bush's ear and he said, "Mr. President, a second aircraft has hit the World's Trade Center. America is under attack."

AP national photographer Amy Sancetta had just bought a pair of super-fast Nikon DIH cameras to cover her 10th U.S. Open tennis tournament. When her phone rang it was the AP desk telling her a plane hit one of the WTC towers and she took a cab and rode down Broadway until a police barricade stopped her. By then, the second tower was smoking. She got out her 80-200 mm zoom lens camera ad as she heard a thunderous rumbling saw through her lens the top of the tower cracking and falling in on itself and got about a half-dozen photos before the building disappeared. Then took her other camera with its 14 mm, wide-angle lens and began taking picture. People running from a gray dust cloud coming toward her bumped into her. Just as the gray cloud neared she dashed into a parking garage as this cloud whooshed past. When she emerged it was like a winter wonderland of gray dust. As she walked toward the trade center, shooting pictures she heard a second rumble and turned and ran until she reached the AP office as the North Tower collapsed.

Gulnara Samoilova is a late sleeper but the wail of sirens woke her. Turning on the TV at 9:03 she saw on TV the *second plane* hit the second tower. As she was four blocks from the WTC she grabbed her camera and film, and headed to the WTC.

Entering the south tower she decided the scene was too chaotic and retreated. Outside, she was standing right beneath the south tower, its smoking bulk filling her 85 mm lens when she saw the tower begin to crumble. She got off one more shot before someone screamed, "RUN!" The force of the collapse "was like a mini-earthquake, and knocked her down and people trampled on her. She got up and as the cloud was about to envelop her, she dove behind a car and ducked down. She said it was like a strong wind and her eyes, mouth, nose and ears were covered with sand, dust, and dirt. "It was dark and silent," she says. "I thought I was buried alive."

Then she heard the fluttering of thousands of pieces of paper. Her sight returned. She had survived. She changed film and lenses and looking down Fulton Street she saw survivors staggering out of the debris. She stepped from behind the car and began taking pictures. A powerful image emerged of a man holding a jacket over his mouth, while a woman next to him brushes debris out of her hair. "I love that photo," Samoilova says. "It looks like a sculpture."

She took her pictures in black and white and when people asked if she wished her photos had been in color, she would reply, "It wouldn't have mattered," Samoilova replied, "Everything was covered in dust - gray dust."-

After September 11 and returning to her job in the AP library, she had to archive AP's photos and cried so much left AP in 2003 and started her own photo studio.

‘   ‘   ‘

The men and women trapped in the top floor of both the North and South Towers left telephone messages to the people they cared about the most o say goodbye. If no one answered they left messages to say goodbye/

Some held hands with colleagues, some wrapped their arms around those they had worked with and stepped out into the empty space. Some went alone. None could understand why no one came to save them. No one understood the fire. Their last words and their voices are unforgettable. One man told his brother, "I'm sorry we fought. I love you."

Another left a message to his mother: "I love you mom. I'm sorry."

Did anyone understand what happened? A plane, an explosion, and why no one came to save them? If the office phones were not working, they used their cellphones. Towers had helicopter landing sites on the roof and many rushed to the stairway to the roof only to find the doors locked. Those who got to the top found the roof under their feet was too hot and started to jump. One said, "I have to go," hung up and jumped. For weeks after 9/11, New York City's radio news coverage played the recordings.

Workers from buildings on Broadway and Liberty, and Chase Manhattan stood away from the towers and watched people dropping in front of them. Tom prayed and hoped his cousin would appear. He stood next to a man who counted out loud each time a body hit the pavement, "25, 27, 28," But Tom's cousin never showed.

On the roof they waited as long as they could until the heat was so unbearably hot they called loved ones, apologized for taking their own lives, said goodbye and jumped. But the heat did not come the explosions in the North Tower. One woman is even shown looking out from the opening hole made by the explosions. People passed by on one stairway to get to the ground and the fires making the black smoke were from oxygen-starved fires. The heat came from the explosions in the basement from thermite and mini-nukes.

## Fort Carson Soldiers on a Killing Spree
### by Dan Edge of *This World*

Seventeen US soldiers from a Colorado military base have been linked to violent killings and attempted killings since their return to US soil. Three of them came from one platoon - highlighting how a generation of American soldiers is struggling to cope with life after military combat.

"I was having a total mental breakdown," one has said. "Every day we were going into battles and never having a break - it was just crazy."

Four members of the Third Platoon are now in prison after serving in Iraq. "I just got to where I couldn't take it. I tried to go to mental health people and they put me on all kinds of meds, too. I was still going out on missions with different medications, different doses, and nothing worked."

Kenny Eastridge is a decorated gunner but is now serving 10 years in prison for his role in the murder of fellow soldier, Kevin Shields, in Colorado Springs. In November 2007, Kenny and two other soldiers, Louis Bressler and Bruce Bastien, were drinking with Kevin after returning from a rough combat tour in Baghdad. Drunk and stoned, they drove off to find more alcohol. Minutes later, Specialist Kevin Shields lay dead, gunned down in a drunken argument, and left by the side of the road in a pool of blood.

Bressler and Bastien were sentenced to 60 years in prison for the murder and a string of other crimes in Colorado Springs. However, the murder of Kevin Shields' was not unique. At Fort Carson military base, 17 soldiers have been charged or convicted of murder, attempted murder or manslaughter in the past four years.

Ryan Krebbs, Third Platoon medic made this statement about military combat and its effect on the lives and minds of men.  "The first six months you're just happy to be home.  Then after that... problems started"

For over a year, This World has been tracking members of Third Platoon, Charlie Company, 1st battalion, 506th infantry (now the 2nd battalion), 12th infantry regiment and effect of combat on them.  The majority of Third Platoon served multiple combat tours with distinction and managed to adjust to life after Iraq but a significant minority has not adjusted.  Four of the platoon members have ended up in prison. Two ended up dead - one from an overdose and the other by a suicide bomb.

In all, 15 out of 42 soldiers from Third Platoon left the army after a single Iraq tour.  Four were kicked out for failing drug tests, one was sent to prison for driving while drunk and fleeing the scene of an accident, and five were medically discharged.  Only five left the army because their service had ended.  More than half the platoon said they suffered from psychological problems after Iraq.

The platoon's youngest member, Jose Barco, is serving 52 years for shooting and wounding a pregnant woman when he opened fire at a party in Colorado Springs.  Barco said he became desensitized to death and killing during the *surge in 2007*, when his battalion was driving al-Qaeda out of Baghdad.

It was Third Platoon's job to move mutilated bodies every morning and he said, "It got to the point where it was like seeing a dead dog or a dead cat.  If you're not numb in those things you're going to go crazy.  It just follows me,"

Third Platoon's discipline deteriorated.  Jose said that brutality was the norm.  He routinely shot unarmed Iraqis.  "We were trigger happy. We'd open up on anything. They even didn't have to be armed. We even kept scores."

The US army investigated. Undoubtedly there were civilians killed by American soldiers, but no one from Third Platoon has been charged with killing civilians in Iraq.

While in Iraq, Kenny Eastridge had exhibited signs of post-traumatic stress disorder (PTSD) and was taking anti-depressants and sleeping pills and also simultaneously taking valium, smoking pot and drinking whisky. Eastridge had a history of aggression, and charged with assault before he went on his second tour, but he was still deployed.

Eastridge said Iraqi civilian deaths didn't bother him "You disassociate yourself. To you, they're not even people, you know. Like, they're not humans.

• August 2004 - August 2005, the Third Platoon was stationed in the desert in the heart of the Sunni triangle to patrol the highway from Ramadi to Falluja to draw out insurgents and dealing with 1,000 Improvised Explosives Devices (IEDs)

• September 2006 - December 2007. Stationed near Al Dora, known as an al-Qaeda hub that became the site of al-Qaeda's last stand in Baghdad. Main task was secure neighborhoods street by street to halt the widespread sectarian killings. The platoon's first battalion commander, Colonel David Clark, accepted the fact that the experience of battle can affect the men. "It's bound to have an impact," he said.

Dan Edge asked, "Is that a reason not to do the surge?"

"No," Colonel Clark replied. "The surge worked. We needed to do the surge. War is a dangerous thing."

The number of Fort Carson soldiers failing drug tests rose by 3000% in the first three years of the Iraq war. Ryan Krebbs, the platoon medic, admitted he used sleeping pills to calm himself down after missions. He never forgave himself for the death of one of his sergeants and attempted suicide by overdosing with anti-psychotic after returning home.

143

"In the first six months you're just happy to be home. Then after that... problems started." Krebbs continued, "Depression, anxiety, paranoia, getting the feeling that you're in Iraq all over again, I just couldn't take it anymore," he said.

Dr. Joseph Glenmullen, Psychiatrist at Harvard Medical School warned, "All of these anti-depressants now carry a warning that anti-depressants can make people suicidal. Side effects are insomnia, anxiety, agitation, irritability, hostility, impulsiveness, and aggressiveness all of which affect troop morale and conduct in the military and out.

The vice-chief of the US army, General Peter Chiarelli defended the policy saying, "It's a supply and demand problem. I cannot do anything about the demand. I have only a finite supply. When the demand goes up and orders are given, we provide the soldiers."

The army's medical command last year conducted an investigation in to the violence and found that soldiers who had experienced unusually intense combat in Iraq, six of them had criminal records before they joined the military, 11 had a history of substance abuse and nine were taking psychiatric medications. Before the Iraq war, American soldiers on psychiatric medications were not allowed to deploy to a combat zone. But by the time of the surge, more than 20,000 US troops in Afghanistan and Iraq were taking anti-depressants and sleeping pills. The military has come under fire for medicating soldiers rather than taking them out of combat. The report concluded that the intensity of battle and shortcomings in mental health treatment have resulted in alcohol and drug abuse but did not go so far as to say it was related to PTSD.

The final American combat brigade pulled out of Iraq after more than seven years of war. But for many the end of combat is just the beginning of a different struggle back home is

144

the consequence of removing guilt and remorse from the minds of men in training them to kill. In an article on the suicide of a female helicopter pilot appeared on March 7, 2015, it quoted a statement from the US Department of Veteran's Affairs that on an average, 22 veterans of combat in Iraq and Afghanistan commit suicide every day as a result of PTSD (Post Traumatic Stress Disorder). Not mentioned are the physical disorders suffered by military personnel exposed to radiation from *depleted uranium in ordinance,* vaccines, chemical weapons such as *agent orange or sarin gas used in crowd control, and biological weapons.* Between 221,00 and 440,000 veterans are disabled permanently or partially in addition affected with PTSD. Mr Bush and Mr. Cheney should be tried in the United States for these are crimes on servicemen.

## News Media Influence on Norwegian Murders

On July 26, 2012 at 2:09 PM on the day of the attack, Anders Behring Breivik emailed to 1,003 recipients as if coming from a Western European Patriot with the message "It is a gift to you... I ask you to distribute it to everyone you know," and less than 90 minutes a bomb made by Breivik exploded outside a government buildings in Oslo. Breivik e-mailed his document to 250 British contacts which doesn't say anything about their relationship to Breivik.

Psychiatric and psychological examinations of Breivik began around five years of age as his father who had divorced his mother attempted to gain control of his son. But the courts ruled in favor of the mother. Part of the father's legal battles was to have psychological tests on Breivik and his mother which revealed his mother had mental problems about male figures.

One psychologist noticed Breivik's smirk but didn't evaluate it. Breivik had plastic surgery done on his face and was very satisfied with results.

Breivik's father moved to France and remarried when Breivik was just a child. He would go to France to be with his father and step-mother during his summer breaks but return to Norway for his schooling.

Anders Breivik was first diagnosed as a psychopath which he knew would commit him to prison for life, he protested, so a second panel was called and labeled him a narcissist which meant anyone praising him would thus be able to control him but no such group or person has been found. Thus his ideas had to have have come from news media hammering on Islam as a religion that incites Muslims to terrorism.

Norwegian forensic experts examined his computer and said he was influenced by fascist groups. One of these groups was SIOE (Stop Islamification of Europe). Breivik said: "I have on some occasions had discussions with SIOE and EDL and recommended them to use certain strategies." Breivik boasted he influenced them which should have him diagnosed as a manipulator (sociopath) and not a narcissist. Five days after Breivik's murder spree, Norway's domestic intelligence Chief Janne Kristiansen, told BBC, "So far we don't have evidence of cells in Britain or in Norway."

Breivikmet with right-wing extremists in 2002 and they "reformed" the Knights Templar Europe and "seize political and military control of western European countries and implement a cultural conservative political agenda". The meeting was attended by representatives from France, Germany, the Netherlands, Greece, Russia, and Serbia.

Paul Ray living in Malta was a founding father of the EDL and denied he ever had any contact with Breivik. He says, "It does worry me that he got inspiration from my blog." Ray has met with Nick Greger, a German terrorist and Johnny 'Mad Dog' Adair, the former Ulster Freedom Fighters Brigadier".

September 11, 2009, SIOE co-sponsored a demonstration with the English Defense League (EDL) in Harrow, UK, to demonstrate against the building of a Harrow Central Mosque. SIOE asked the Jewish Community for 1,000 Jews to carry the Israeli flag to support the demonstration.

Rabbis Frank Smith, Aaron Goldstein, Hillel Robles, Mike Hilton, and Rabbi Kathleen Middleton sent a letter saying, "As leaders of the Jewish community in Harrow, we are writing to express our support for our Muslim friends and neighbors ... who are under attack from those whose only purpose is to spread hatred and fear.

SIOE was originally a Danish anti-Islamist group which developed out of the Jyllands-Posten Muhammad Cartoons. In 2007, Anders Gravers Pedersen founded the small Danish political party called the Stop the *Islamification* of Denmark. On 11 September 2007, the group staged a demonstration in Brussels, Belgium (NATO HQ)

Affiliate organizations have been created in Russia, Finland, France, Germany, Norway, Poland, Romania, and Sweden, as well as the United States of America. But think of the popular Iraqi couple killed in America by their neighbor who was an atheist and two shootings of Christians in South Carolina and Oregon in October 2015?

If Israelis and Americans did 9/11 what would be the reaction of Islamic radical? All the evidence correlates with Dimitri Khalezov's revelation that Mossad Colonel Mike Harrari was the mastermind. How will those who preach Islamophobia people and anti-Muslim react? Well they will deny this. They are firmly convinced they are right. The biggest change will be in Israel and it is already starting.

## This is Where ISIS Began

On July 31, 2009, three young American Jews were captured after crossing the border from Iraq to Iran in the Zagros Mountains of northern Iraq. They said they were hiking to visit a waterfall in the mountains for a picnic. Sarah Shourd, 32, was teaching English in Damascus and helping refugees displaced by the rebel fighting in Syria. She and her boyfriend, Shane Bauer, who speaks fluent Arabic and is a free-lance journalist who had worked for *Mother Jones, L.A. Times. Democracy Now, The Nation,* and *The Christian Science Monitor*, were joined by their friend, Joahua Fattal, whose father emigrated from Israel to the US and has dual Israeli-American citizenship. The three young Americans from California went hiking on the Iran-Iraq border, caught on the Iranian side charged with spying and held in detention.

After a year in prison, Shourd began screaming and smashing her fists against the wall of her cell, The Sultan of Oman (that has given Muamar Gadhafi's daughter, Ayesha, sanctuary) and he paid $465,000 to Iran for Shourd's release on humanitarian grounds. Fattal and Bauer were tried and sentenced to 8 years for espionage. But after 4 years they were released with $465,000 each paid by Oman's Sultan.

How did these three Cal-Berkley graduates get from Damascus to the Zagros Mountains on the border of Iraq-Iran? This area was where the Mossad stirred up Iraqi weres to attack Saddam Hussein. But when weapons Israel supplied the Kurds showed up in Turkey and were used to kill Turkish soldiers, Turkey complained and Israel pulled the Mossad out of Iraq.

This Kurdish area of Iraq next to the Zagros Mountains was the base area from which this ISIS has suddenly emerged to overwhelm and capture Iraqi military bases. After Iraqi soldiers surrendered to the ISIS they were then beheaded by this ISIS whose warriors began kidnapping and raping women and beheading hundreds of not thousands of a religious minority, the Yazidis who lived around Mosul.

The author first learned about the Yezidis from a documentary back in 1972 and asked an Iraqi friend, Qais Al Shabbat, if they really worshipped Satan. Qais replied that they believed in a god that was both good and evil whose name they were forbidden to speak and Muslims and never bothered them.

ISIS identifies itself as Sunni and attack Shiites who support Bashar Assad who is Sunni. ISIS uses psychological warfare techniques to divide Shiites and Sunnis in Lebanon, Syria, and Iraq to organize a united front to oppose them.

The logistics of bringing thousands of Western converts to Islam, food, weapons, and finance while keeping all secret until bursting on the scene is something that CIA, British Intelligence, and French Intelligence should have been aware of with satellite monitoring and Iraq government would have learned about. The logistics in transporting these men to Iraq, food, weapons, uniforms, training, and housing would be so hard to conceal. So how large is ISIS?

Iran's news agency has reiterated that Israel's Zionists created ISIS from the previously planned Islamic State Iraq and the Levant. MI6, the CIA, and Israeli intelligence resurrected the *Hornet's Nest* from a prior Israeli-British operation called PLAN Britannique.

An independent American journalist, James Foley went missing in November 2012. The ISIS videotaped his murder by ISIS *Jihadi John* dressed in black who cut Foley's throat cut and then severed his head from his body. This was followed with the beheading of American Steven Sotlof by an ISIS murderer. Both Foley and Steven Sotloff were dressed in orange suits like Muslims detained at Guantanamo as were the captured Iraqi soldiers who were beheaded by ISIS because the detainees at Guantanamo have to wear these orange colored jumpsuits. The beheading by ISIS was taught by the CIA in Latin America and Afghanistan as counter-insurgency which uses this as psychological warfare. .

The *Somdaily News* reported that Israel came up with this idea of creating ISIS as an enemy near its borders to get support of oil exporting nations to come the aid of Israel. By assembling all the facts, Yasimina Haifi of the Netherlands Security office has said the US and British policy to eliminate Saddam Hussein was to create a political vacuum which is behind the Israeli creation of ISIS as a threat near to Israel that will result in their expansion of Israel to make a Greater Israel.

The project manager of the Dutch National Cyber Security Center, Yazmina Haifa, Tweeted that El Baghdadi, the leader of this ISIS or ISIL is a Jew trained by the Mossad and the so-called Islamic State, ISIS beheading children and captured Iraqi soldiers "has nothing to do with Islam. It is a preconceived plan of Zionists who to make Islam look bad,"

Abu El Baghdadi who poses as an imam pure Islam started his movement in 2006 about the same time as the three American Jews were captured inside the border of Iran who came all the way from Damascus to the Zagros Mountains on the Iraq-Iran border to go hiking. The murderous ISIS began attacking the

Yazidi minority of Iraq and then it was everybody the imams would say who to murder and they did it.

Yazmina Haifi was suspended from her management job in the Ministry of Security and Justice. The National Coordinator for Counterterrorism and Security (NCTV said, "Security and Justice and the NCTV distance themselves from her remarks. Since her comment relates to the work of the NCTV and the National Cyber Security Center, cause is shown to terminate her assignment NCSC/NCTV and outsource Yasmina Haifi's work with immediate effect," (NL Times).

Yasmina Haifi on Netherlands local Radio 1, Yasmina said "Freedom of expression is apparently only for certain groups. I have taken the liberty to express myself and obviously I have to pay for it. I do not know why I should take it down; what I think." She refused to retract her statement. Netizens that controls Twitter adds Haifi removed her comment from Twitter and written, "I realize the political sensitivity in relation to my work. This reaction was never my intention."

Two Dutch MPs, Joram van Klaveren and Louis Bontes, asked the Security and Justice Minister, Ivo Opstelten, to punish Haifi and said in a joint statement"...a person who propagates such ideas and works with state secrets could be a threat to national security." Haifi was transferred to another position.

*Veterans Today* offered translations from Radio Ajyal.com and the Egyptian Press that the leader and Caliph of ISIS, *Abu Bak Al Baghdadi is* a Jew whose real name is Simon Elliot and trained in psychological warfare by the Mossad in the Negev. Accordingly, the plan is for ISIS to get into the military and civilian heart of the Middle East oil fields and facilitate a takeover to make a Greater Israel. As Israel has never defined its borders this could include Iran's Khuzestan where the Prophet Daniel is located, Iraq, Kuwait, and the Persian Gulf to Egypt where Joseph is buried.

According to Yahoo.com, the ISIS has attracted Muslim Converts from Western Countries as warriors. Are they mercenaries or are they converts who were so frustrated, abused, threatened, and persecuted to have flipped into becoming fanatics? The British used an Arab Muslim working for them to get the Muslim brotherhood to fight Nasser in h3 1960's by saying Nasser was doing the work of Satan by *declaring Egypt a secular nation*. Nasser was preoccupied in a war in Yemen for six years allowing Israel to make its Six Day War (1967) on all its neighbors and attack the USS Liberty to blame on Egypt and have American support for Israel to annex the Sinai, Gaza, West Bank, and Golan Heights.

Nasser made friends with the CIA and in 1968 introduced Saddam Hussein to the CIA when he was sent to Nasser with the proposal to unite Baath party of Iraq. Muamar Gadaffi also proposed uniting Libya with Egypt that would give Egypt oil as well as money to finance development projects. All of this panicked Israel and they poisoned Nasser in 1969 and put the blame on the blind cleric, Imam Omar Rahman, who ws jailed and later cleared and released. When the CIA asked MI6 to use the Muslim Brotherhood to recruit Mujahedeen to fight the Russians in Afghanistan, they gave him Omar Rahman.

Anwar Sadat, was a German agent in WW II and changed to be a British agent, succeeded Nasser. Sadat began creating a coalition to attack Israel and return the Middle East to the borders to the 1967 borders that became the Yom Kipper War in 1972. Israel suffered heavy losses in the beginning but billions in American military weapons were shipped to Israel and the US provided satellite photos showing the location of the Arab Armies shifted the war to Israel's favor.  But Arab oil producers imposed an oil embargo on the US for helping Israel.

Were these Jihadists targeted by police in England and the US frustrated, fired from jobs, and persecuted until they snapped?  Controlling every aspect of a man's life by spying on and frustrating will put him into mental state where he'll crack, get involved with ISIS, or suicide bombing.

Edward Snowden revealed that NSA sent information on emails, phone calls, faxes, cables on Arabs and Palestinians living in America who have relatives living in Palestine and Israel to Israel's IDF Unit 8200.  The information sent also contained sexual orientation, infidelities, money problems, family medical problems, or other problems.  43 veterans of the IDF Unit 8200 complained that this abused innocent Palestinians by pressurizing them to spy for Israel and create divisiveness.  Israel says it will investigate.

The secret society of Betar was a formed in Riga, Latvia in 1924 with the goal that Israel have both sides of the Jordan River and all the Middle East.  They killed 3,000 ethnic Germans in Czechoslovakia and 9,000 ethnic Germans in Poland to get Hitler to invade and Britain used the League of Nations to declare wron Germany.  The Betar  joined Haganah, Irgun, and Stern gangs in killing hundreds of British soldiers and PalestinOians before independence in 1948 that pushed the British into withdrawing from Palestine and led to Israel's war of independence against Palestinians who had not protection.

A German Jewish couple posing as arms merchants sold defective weapons to Egypt's King Farouk.  Rifles blew up in the face of Egyptian soldier, tanks ground to a halt with sand in their oil. Colonel Abdel Gamal Nasser with other colonels overthrew Farouk and put General Naguib in as president.  But Naguib was being advised by Abdullah al Alawi, who was coerced to make

Arabic language broadcasts for the Japanese in WW II and would be executed for his war crime. He ran away to Indonesia and made a deal to work British.Intelligence.

Abdullah's father-in-law was Ismail Badawi, the youngest French General of WW I and a descendant of the Prophet Mohammed. He commanded thousands of Moroccan and Senegalese troops fighting for France in WW I. After Syria was mandated to France by the League of nations, he was sent to Syria to make it more French.

Abdullah was coerced by the Japanese in WW II to make Arabic language broadcasts from Singapore directed at Arabs in the Indian Ocean. Abdullah was to be executed by the British as a ar Criminal and to escape execution, he became a British agent and assigned as an advisor to General Naguib, who was installed as ruler by the Egyptian colonels after overthrowing King Farouk. But as Naguib was making pro-British decisions, the colonels threw him out and installed Abdel Gamal Nassed.

So Abdullah used the Muslim Brotherhood (which was controlled by British Intelligence, to stage a mass rally for Naguib in downtown Cairo, and Nasser putting him back in power. Then an attempt on Nasser's life stage the removal of Naguib for ever and Nasser called for a United Arab Republic which Israel saw as threat with Syria and Yemen joining after a bloodless coup overthrew its monarchy with a republic that had a Constitution protecting the rights of the people.

Yemeni officers went to Egypt for military training where a coup orchestrated by the British and backed by the Shah of Iran, the Gulf Emirates, and Saudi Arabia aiding.

Israel began immediately to force Muslims from their homes and confiscate their properties without compensation during Israel's War of Independence.. A campaign of ethnic cleansing and massacres of Palestinians followed. Palestinians had no government or army having lived under the protection of Turkey for centuries and then under British Mandate from the League of Nations and the United Nations did nothing.

Not all Jews are terrorists but in 1938-39 the Betar terrorists killed 2,000 ethnic Germans in Czechoslovakia and 9,000 in Poland to push Hitler into invading both countries. Churchill used the League of Nations to declare war on Germany. All nations in the League of Nations were to contribute men and arms. But as the Constitution states that the chief of all the armed forces was in the President and only Congress could declare war, the United States did not join.

A German Jewish couple posing as arms merchants sold defective weapons to Egypt's King Farouk. Rifles blew up in the face of Egyptian soldier, tanks ground to a halt with sand in their oil. Colonel Abdel Gamal Nasser with other colonels overthrew Farouk and put General Naguib in as president. But Naguib was being advised by Abdullah al Alawi, who made Arabic language broadcasts for the Japanese in WW II and then forced to be a agent or be executed for his war crime.

As Farouk bought defective weapons from this German Jewish couple, the Egyptian colonels overthrew King Farouk and put General Naguib in as leader of Egypt and Abdullah was to be his adviser. But Naguib made pro-British decisions so the Egyptian colonels replaced him with Abdel Gamal Nasser. Abdullah used the Muslim Brotherhood to stage a mass rally of thousands in downtown Cairo demanding Naguib be returned to power, Nasser acceded. Then an attempt on Nasser's life a week later was blamed on Naguib, and Nasser returned to power and called for a United Arab Republic.

As all decisions on governing a tribe or nation had to be decided by the people in a vote. Hereditary rule in is forbidden. After the Prophet passed away, a Caliphate was selected by vote and his power was centered in Baghdad until the Ottoman Turks took Constantinople and transferred that power to Istanbul and the Ottoman Sultan became head of Islam.

So a bloodless coup overthrew Yemen's monarchy and it became a republic and it military officers went to Egypt for training when the Yemen Republic joined Nasser's United Arb Republic. But a counter-coup orchestrated by the British with support by the Shah of Iran, Gulf Emirates, and Saudi Arabia aiding, an Imam was hustled in and installed (monarchy). 21 men were beheaded including Abdullah al Wazir who wrote Yemen's Constitution were beheaded and a.ll the children and family of those executed were taken to fortress in the mountains where they were kept for five yers until the Yemeni Tribesmen surrounded the castle, show it up and liberated everyone.. Among them was Ibrahim Abdullah al Wazir, who chose to study at the world's Al Ahzar university, established on books from the Ancient Library of Alexandria. Ibrahim received a Masters and was chosen to lead the last two tribes of Yemen.

So now Abdullah is directed to be the adviser to Ibrahim and Nasser wanting to bring back Yemen into the United Arab Republic, Abdullah tells Ibrahim, "Nasser is doing the work of Satan by call Egypt a secular nation instead of an Islamic nation.

So the Yemen tribes take on the Yemen Army and Nasser invades. So for six years, Nasser is tied down in a war with Yemen's tribes and not paying attention to his 15,000 troops in Gaza. Israeli planes flying low over the Mediterranean turn at the Libyan-Egyptian border and fly toward Cairo. Egyptian radar pick technicians think it is American planes from Wheeler Air Base and do not give an alarm. 300 top Russian planes are destroyed and 15,000 Egyptian soldiers put up little fight with the invasion of Gaza, where 1,000 were executed by Israel (a war crime) with some having their hands nailed to palm trees and used as targets.

156

Orders for the USS Liberty to leave the Mediterranean were sent to the Philippines wile orders from Washington direct it to 17 miles from Gaza and 12 ½ miles from Sinai, now all occupied by Israel.

The author added this information to show: 1.The Muslim Brotherhood was controlled by British Intelligence (and labeled a terrorist organization). 2: The British use of Islam to influence division. 3. The use of all the diverse minorities in every nation in the Middle East to cause rebellion.

a result of 9/11 and caused some Muslims to do rash things. While Israel justifies is use of military weapons against civilians as *Israel's right to defend itself,* its detention of thousands of Palestinians, and its pushing the United States to make regime change in Iraq, Afghanistan, and Libya, has cost many lives, wasted money and resources ultimately parallels the PNAC. Both PNAC and Israel oppose an autonomous international investigation of 9/11.

This is the other side of the Israeli attack on the USS Liberty on June 11, 1967. Ibrahim is a good man and loves his people and believes in Islam and God. He resigned as leader of the tribes and survived an assassination attempt in a suburb of Detroit and was poisoned in his villa in Jeddah but survived

This book documents the financial and international policies that 911 has affected NATO, the US and allied nations, Russia, the People Republic of China, the two Koreas, and the United Nations Assembly. The interrogators of detainees at Guantanamo, Bagram, Bahrain,s Turkmenistan, Kyrgyzstan, Afghanistan, and Morocco were trained by IMS (Information Management Systems) which was created and owned by Lockheed Martin, a main weapons supplier to the US Air Force. While IMS has claimed great success in getting information on future terrorism using torture and threats, Congress investigated and found *no useful intelligence was obtained.*

The NIST report was started one year after 9/11. President George W. Bush gave out unsupported allegations and delayed in investigating while covering up the use of torture on detainees and kidnapping by the US throughout the world.

The author defines torture as being physically restrained to endure pain administered by someone mentally unbalanced.

But there are other devices used by the US Government and secret societies such a blacklisting, a program of slander that is definitely coming from narcissistic personalities, or murder by assassination units of the CIA that hires people with anti-personality disorders that have no remorse or guilt to silence American *whistleblowers.*

## Men George W. Bush Praised and Promoted

As US Ambassador to the UN Security Council, John Dimitri Negroponte in a closed meeting on July 26,2002 said the United States will oppose any resolution passed on to the 15 nation UN Security Council any condemnation of Israel by the UN General Assembly concerning the Israeli-Palestinian Conflict that does not condemn terrorist groups and must have the four following elements to be accepted:

1. A strong and explicit condemnation of all terrorism or Incitement to terrorism.
2. A condemnation by name of al-Aqse Martyrs' Brigade, Islamic Jihad, Hamas, or groups that have claimed responsibility for suicide attacks on Israel.
3. An appeal to all parties for a political settlement.
4. A demand for improvement of the security situation as

a condition for a withdrawal of Israeli Armed Forces to positions held before September 2000 start of Second Intifada.

Like 9/11, the crimes of murder committed by Israeli Defense Forces in occupied Palestine are never investigated. Killers are never punished such as the murder of an eight year old Palestinian girl who passed by group of a Israeli soldiers on her way to school. They started shooting at her and she ran for cover. The Israelis conversations were recorded: "The target shows fear. She is hiding... " and it goes on until she is dead. When Israel receives a complaint, it says it will investigate.

A video of a Palestinian father shielding his small daughter with his body shows them being shot at by Israeli soldiers. The father took shelter behind a concrete wall and was waving his hand and yelling to stop shooting. When they stop, the child is dead and the father wounded.

Dimitri Negroponte was Ambassador to Honduras where many crimes were committed and he *knew nothing about them*. President Bush appointe him to the UN Security Council where he initiated the Negroponte Doctrine, which is that any crimes charged against Israel must include a condemnation of *al-Aqse Martyrs Brigade*, Islamic Jihad, Hamas, or a groups alleged to have claimed responsibility for suicide attacks on Israelis.

The Second Oslo Peace Accords began in 1995 and a young Israeli settler in Occupied Palestine murders Prime Minister Yitzak Rabin who with Shimon Peres signed on the Oslo Peace Accords for Israel.

John Dimitri Negroponte was born in London, England on July 21, 1939. His father was a Greek shipping magnate. Negroponte attended Buckley School in New York City then went on to Phillips Exeter Academy in New Hampshire graduating in 1956 and entered Yale University where he joined Psi Upsilon Fraternity. Two of his fraternity brothers were William H.T. Bush, (uncle to George W. Bush) and Peter Goss (George's Director of the CIA 2005-2006.  He also joined the *Skull and Bones* secret society at Yale and graduated in 1960.  He enrolled in Harvard's Law School but dropped out to join State Department's Foreign Service.

In 1964, Negroponte was sent to the US Embassy in Saigon as a political officer and learned Vietnamese.  In 1973, he was in charge of Richard Nixon's 1973 secret negotiations for the withdrawal of American troops from Vietnam.  Nixon had agreed to finance the reconstruction of Vietnam in return for the return for about 30,000 America POW's held by the Vietnamese but Fidel Castro told the Vietnamese to hold back five to ten thousand Americans in case America reneged on this promise.  So when America did not rebuild Vietnam you have thousands of Missing in Action families angry their sons were kept behind.

Negroponte's roommate in Vietnam was Richard Holbrooke, Clinton's envoy to Bosnia.  Holbrooke made a secret deal to divide Bosnia between Croatia's Franjo Tudjman and Serbia's Slobodan Milosevic July 15, 1995 five months before Dayton Peace Accords.  French TV Channel 1 and Senator Bob Dole said *Clinton sold out Bosnia*.

After President Nixon appointed Dr. Henry Kissinger as Secretary of State, Henry sent Negroponte to the US Embassy in Quito, Ecuador as Economics Officer and Political Officer.

On August 10, 1979, President Jaime Roldos Aguilera became first democratically elected president of Ecuador and instituted a 40 hour work week and a minimum wage of $160 a month.  He asked the foreign oil companies for a larger share of the oil profits.  Roldos felt the American opposition to the Sandinistas was unjust and allied himself to the Sandinistas.  As the Sandinistas sympathized with the rebels in El Salvador, Roldos allied himself to them as well.

Ronald Reagan was inaugurated President on January 20, 1981.  His Vice President was George H. W. Bush, head of the CIA under President Gerald Ford.  Four months later on May 24, 1981, the plane carrying President Roldos, his wife, Minister of Defense, and others exploded over the jungles of eastern Ecuador killing everyone.  Indigenous people led the investigators to the crash site as they had seen the plane explode in mid-air.

In 1979, President Jimmy Carter appointed Jack Binns US Ambassador to Honduras.  Binns learned of atrocities in neighboring El Salvador by Salvadoran military and sent reports of human rights abuses and murders..  In July, 1981, his superior, Thomas Enders, Assistant Secretary of State for Inter-American Affairs, ordered him to stop using official channels to report human rights violations as these could be seen by Congress and lead to a cutoff of military aid to El Salvador and the Honduran.  The CIA station in Honduras was ordered to withhold information on human rights violations from Binns.

On October 10, 1981, President Ronald Reagan dismissed Binns and appointed John Negroponte as US Ambassador to Honduras.  During his ambassadorship, military aid to Honduras grew from $4 million to $77.4 million a year, and the US increased its military presence there.  President Reagan asked Congress to finance a mercenary army of ex-Somoza National Guard.  Honduran Army Major Zuniga was sent to testify before the Senate Foreign Relations Committee in Washington.  Zuniga admired the Contras and as the CIA gave the Contras a manual that tells the Contras to engage in terrorist activities to destabilize the Sandinista Government, he thought it was OK to tell the Committee about atrocities.  But Congress was horrified and refused to finance the Contras.

On his return to Honduras, Major Zuniga was put in detention by the Honduran Army and then kicked out of the army.

He talked about going into business with a Cuban and Zuniga gave him $40,000. But the Cuban decided to keep the money and hired two Nicaraguans to kill Zuniga. They beat him, shot him in the head, pushed a metal file through his heart and threw his body off a cliff.

In El Salvador, Archbishop Romero was shot in the heart as he says Mass for thousands of Salvadorans standing outside in the street. Then four nuns escape from El Salvador to Tegucigalpa, Honduras, where agents of DIN take them to DIN headquarters and they are tortured and raped. DIN asked El Salvador to send a helicopter to take them back to El Salvador. Loaded on the helicopter, they are pushed from the helicopter as it flies over the Pacific Ocean.

*Operation Charly* was the plan to create a Latin American army to invade El Salvador, drive the Salvadoran rebels across Honduras into Nicaragua where they were to be annihilated. Argentina was to supply most of this army. Nicaraguan Sandinistas kidnapped an Argentine sent to Costa Rica to teach destabilization and guerrilla tactics and put him on TV in Managua. *Operation Charly* was called off.

In September1992, the United States and Nicaragua agreed to have their case heard before the International Court of Justice. When the court decided in favor of Nicaragua and ordered the United States to pay Nicaragua six billion in damages the US used its influence in the UN and paid nothing.

From 1985 to 1987, Negroponte was Assistant Secretary of State for Oceans and International Environmental and Scientific Affairs for President Reagan, then Deputy Assistant to President Reagan for National Security. From 1987 to 1989 was under Colin Powel, Secretary of State and involved in the removal of General Manuel *Pineapple face* Noreiga, who laundered CIA dirty drug money demanded 20% to do it and the CIA was

162

unhappy with that. Mike Harrari, the chief of Mossad operations in Central America was head of Panama's National Guard Air Force and put a huge computer in the Panama National Guard Headquarters which which president- general Manuel Noriega had access to. During the American invasion of Panama in 1989, Panamanian guardsmen took sledge hammers and battered it. US Army Intelligence cut away the twisted metal, took out the hard drive and began accessing how Israel was working against the United States in Latin America. Mike was getting on a plane to fly out of Panama when US Soldiers stopped him and held him until a, phone call to New York City had the soldiers helping Mike board the plane to Israel.

Without money from the cocaine, the Mossad was short of cash to pay off Robert Maxwell, whose lavish lifestyhim into debt with $300,000,000 he owed. When Maxwell threatened to give away some secrets, the Mossad told him to sail his yacht to the Canary Islands. There he anchored his yacht on the others side of the island from Tenerife and waited at night for the three men in a motor powered dingy to give him the money.

Pulling up beside the yacht, Maxwell dropped down a rope ladder. To men climbed up and one seized him while the other put muscle relaxant in a hypodermic syringe to his neck, and before his lungs collapsed, and dropped him in the water to drown. His body was washed up on the beach and that was the end to the Little Czech.

Mike asked for and got 60% of every deal made with

Prime Ministers Yitzak Shamir and Menaghen Begin were Mossad chiefs in Europe and the Mossad sold billions of dollars of drugs in Europe annually. Bernie Cornfield was given fifty million dollars by Meyer Lansky, head of the Jewish Mob to set up Investors Overseas Services to launder Mossad dirty money. It was Lansky who put the Jewish Mob and Sicilian Mafia together to make *La Cosa Nostra.* Meyer Lansky was also the Number One Communist in America in 1952.

From 1989 to 1993, Negroponte was US Ambassador to Mexico where that was a huge increase in drug trafficking. Undercover DEA agent Kiki Camareno was kidnapped by the Mexican cartel in September 1991. They tortured Kikki with an electric Phillips screwdriver to make holes in his skull. A Mexican doctor kept him alive for a week. DEA agents in the pay of the Mexican cartel were present during his torture.

The author pinpointed the location of clandestine drug air strip in Honduras when he was working on economic feasibility studies in Honduras. Juan Ramon Matta used that air strip when he worked for Ollie North in supplying the Contras and flying cocaine into Mena, Arkansas.

From 1993 to 1996 Negroponte was US Ambassador to the Philippines, appointed by President Bill Clinton and ambassador to Greece in 1999. After mass graves were discovered on a former U.S. military base in Honduras where Americans trained Contras during his ambassadorship to Honduras, the Honduran Rights Commission accused Negroponte of human rights violations in 1994. Negroponte could not pass Senate confirmation with the atrocities in Honduras coming to light, so McGraw-Hill made him Executive Vice President for Global Markets.

Three days after 9/11, President George W. Bush appointed him as US Ambassador to the UN Security Council. Senator Christopher Dodd made this observation: "Based upon the Committee's review of State Department and CIA documents, it would seem Ambassador Negroponte knew far more about government perpetrated human rights abuses than he chose to share with the committee in 1989 or in Embassy contributions at the time to annual State Department Human Rights reports."

Senator Dodd cited a 1985 cable sent by Negroponte showing he was aware of the *death squads* of Battalion 316 that carried out assassinations and torture. The battalion was headed by General Gustavo Alvarez Martinez of the Honduran Army who was forcibly removed from his post by fellow Honduran military commanders in 1984.

Negroponte pushed for a reformation of the Honduran Criminal code and justice system after the main power plant in Tegucigalpa was blown up and the kidnapping of the entire business establishment of San Pedro Sula in 1982.

In 1995, the *Baltimore Sun* published an interview with ex-Honduran congressman, Efrain Diaz, who said of senior US officials and Negroponte: *Their attitude was one of tolerance and silence. They used Honduran territory more than they were concerned about innocent people being killed.*

In April 2005, the *Washington Post*, wrote that the image of Negroponte showed he was an exceptionally energetic, action-oriented ambassador whose anti-communist convictions led him to play down human rights abuses in Honduras, the most reliable U.S. ally in the region. There is little in State Department documents released to assert he used *quiet diplomacy* to persuade the Honduran authorities to investigate violations, including the disappearance of dozens of government opponents.

The *New York Times* wrote: *the documents revealed a tough cold warrior who enthusiastically carried out President Ronald Reagan's strategy in Latin America.*

On February 27, 2007, President George W. Bush appointed Negroponte as U.S. Deputy Secretary of State *under* Secretary of State, Condoleezza Rice. He served until the end of the Bush presidency, January 20, 2009.

Ambassador Negroponte serves on the Leadership Council of Concordia (a nonprofit group based in New York City *promoting effective public-private collaboration to create a more prosperous and sustained future.*

Negroponte received the Lifetime Achievement Award by the *World Affairs Councils of America.*

Received the *Jet Trainer Award* for Distinction in the Conduct of Diplomacy, Georgetown University, Institute for the Study of Diplomacy.

*George F. Kennan Award for Distinguished Public Service,* National Committee on American Foreign Policy.

*Honorary Patron of the University Philosophical Society,* Trinity College, Dublin, Ireland.

In 2001, John D. Negroponte instituted his Negroponte Doctrine but was unsuccessful in getting the UN Security Council to sanction an invasion of Iraq in 2003.In 2004, he was appointed US Ambassador to Iraq with L. Paul Bremer the supreme authority. *Newsweek* magazine reported that the Pentagon was considering *calming* Iraq with *Salvadoran Options* which was entering villages and killing people suspected of sympathizing with the rebels fighting in El Salvador. So *death squads* roamed Iraq and mass murders became commonplace like in Honduras.

Another man President George W. Bush promoted was L. Paul Brenner, a long time diplomat in State Department's Foreign Service. He served in Afghanistan, Malawi carrying out embassy functions and then as an assistant to Henry Kissinger under Nixon. Ronald Reagan appointed him to the Netherlands. Bremer retired from State Department in 1989 and became chairman of Marsh and McLennon and then its CEO and was replaced by Cherkasky after George W. Bush appointed Bremer the Administrator of Coalition Forces in Iraq as the *first supreme authority of the coalition forces in Iraq* with *absolute powers* on May 9, 2003. Bremer was to lead the reconstruction and management of Iraq with sales of Iraq's oil. The Inspector

General's first report said 9 billion dollars was missing and asked for interviews with accountants *and* was denied.

Bremer's office in the North Tower was the floor above where *flight 11 supposedly hit.* At 6:45 am, Israel's Mossad sent a message to the CIA to tell employees not to go to Twin Towers. Bremer was to meet his staff in his office at 8:30 am, but Bremer changed the meeting to a nearby hotel. He had fifteen minutes he could have notified them of the change because at 8:45, *Flight 77* hit the four floors of the North Tower and his staff died. A photo of the hole made by Flight 77 has a girl waving from that opening. The four floors occupied by Marsh and McLennon underwent *special modifications* no other floors had.

The official report says Mohammed Atta flew Flight 77 into Marsh and McLennon's offices and took cocaine while living with a DEA agent when he was taking a course on how to fly a Cessna.

Bremer was an outspoken defender of the Iraq war and when Erin Burnett interviewed Bremer on CNN on June 16, 2014, she told him a lot of people would be wanting to hear him explain his side of the controversy surrounding him.

Bremer responded that he had done nothing wrong.

When Erin asked about areas that the ISIS has taken over with the people rejoicing that they once more had electricity, Bremer replied that he had restored electricity to levels above pre-invasion levels. A Wikipedia article states that during Bremer's time as head of the coalition occupation, May 12, 2003 to June 28, 2004. "Iraq's oil infrastructure was repaired rapidly but the reconstruction and repair of Iraq's potable water, sewage, and electricity systems was extremely slow."

Bremer's defenders say this due to the unanticipated fierceness and resistance to the coalition's occupation. Critics blame it on Bremer's preference for US firms as Iraqi firms only

167

got 2% of the reconstruction contracts.  Another criticism was that the CPA's employed by Bremer were unqualified as auditors and never filed reports on time.

1,750 years ago in China a wise and clever Kong Ming defeated Men Ho six times and let him go until Meng Ho submitted the seventh time.  Kong Ming's generals complained that he should have executed Meng Ho the first time, but Kong Ming replied: *"We would have to send administrators to govern their nation and always have to send in soldiers to protect them and they would always hate us.  This way there'll be peace.* Republican Congressman Newt Gingrich who has a doctor's degree in history said: "Bremer was the largest single disaster in American foreign policy in modern diplomacy."

Four star USAF General Richard Myer was chairman of the joint Chiefs of Staff on the morning of 9/11.  He was advisor to President Bush and Secretary of Defense Rumsfeld in the Iraq War.  The morning of 9/11, an exercise drill was conducted by NORAD, Space Command, and all defense agencies on a simulated drill of a small plane being hi-jacked and flown into a tower of the National Reconnaissance Office.  No actual plane was to be involved but exits and stairwells were closed off as if it had been hit.  When Chief of theArmed Forces, General Myers was asked by Congressional Committee why the safeguards to protect America from the 9/11 attacks, his reply was, "We were geared to protect from outside America, not inside America," Vice President Cheney was conducting an exercise to test for such an attack given names like Operation Northern Vigilance, Vigilant Warrior, and Vigilant Guardian.

From 2000 to December 31, 2001, Bernard Kerik was New York City Police Commissioner from 2000-Dec 31, working mainly with the Department of corrections and credited with cutting the crime rate in NYC.  He was appointed Minister of

168

Interior of the Iraq coalition in 2003. When President George W. Bush wanted him to head Homeland Security he declined as he had hired an illegal immigrant as nanny for his children. In 2008, Kerik pled guilty to fraud, lying under oath, criminal conspiracy and sentenced to 4 years in prison on February 13, 2010.

After the 2003 defeat of Iraq, contracts to supply workers for the reconstruction of Iraq were meted out to companies that hired thousands of Contract Workers. American contractors hired skilled workers to work in Afghanistan and Iraq. Americans were the majority but workers from fifty different nations were employed like Fiji, Nepal, Germany, UK, and South Africa. The contractors who hired them were not interested in their protection and health issues. 3,258 died, 89,181 were injured and 1,101 missing between Sept 30, 2001 and Dec 31, 2012. Workers have lost homes, divorces, families split up, and suicides from PTSD (Post Traumatic Stress Disorder) that come with the high pay.

An Illinois truck driver who was making $45k a year would earn around $100K in Iraq or Afghanistan. The most well-known of the civilian contractors in Iraq was Blackwater that supplied mercenaries. Erik Prince was the founder of Blackwater Worldwide and hired thousands of mercenaries to protect contract employees and the projects on which they worked. Mr. Bush got immunity for all Americans from prosecution by the IWCT but four Blackwater security guards were tried in an American court for killing 14 unarmed Iraqi civilians in a Baghdad traffic circle in 2007. One was sentenced to prison for life, three were each sentenced to 30 years in prison.

In 2009, Blackwater's name was changed to Xe. In 2010,, Eric Prince the founder of Blackwater sold it and Xe's name was changed to Academi in 2011. In June 2012, Academi merged with rival firm Triple Canopy to form Constellis Holdings, Inc. On the

board of Constellis' is attorney John Ashcroft. By using mercenaries, that absolves the country hiring them from being charged with War Crimes. It is also
used to cover theft, murder, and more.

## Tayyip Erdogan's Rise to Power

To understand the 2016 coup attempt in Turkey, one has to go back the defeat of Turkey and the division of Ottoman Empire between Britain and France. General Mustafa Kemal Ataturk escaped Istanbul by boat to Samsun on the Black Sea and made his way to Ankara where he mobilized an army that forced the occupation forces out of Anatolia and issued a declaration of Independence for a Turkish Republic. .

Ataturk insisted the conference be held in Switzerland as it was neutral in WW I and he chose the city of Lausanne. The Allies suggested to Ataturk that he make himself sultan but he declined saying he liked his whiskey too much. The Allies demanded that a Turkish military hold the power to diminish any influence of Islam and never attempt to recover the empire

Under the military of Turkey from its inception as a Republic, the Turkish economy was a mess until Ali Adnan Retake Mendares became Prime Minister in May 1950. Adnan was of Crimean Tartar origin and entered politics. In May, 1950, he became Prime Minister. He needed to modernize Turkey and got German banks to accept Turkish lira at 3.5 to the dollar and foreign companies flocked to Istanbul for contracts. Turkey boomed but German banks were overloaded with lira they discounted it to15 lira to a dollar.

President Menderes got Turkey into NATO and even sent a battalion of Turkish troops to join the UN in fighting North Korea. He signed the Bagdad Pact with Iraq and Iran to keep Russia out of the oil rich Middle East. However with Baghdad

Revolution, America rushed thousands of WW II weapons to Iran as the Baghdad Pact existed only on paper. .

The Turkish military arrested Adnan Menderes on May 27, 1960 accusing him of attempting to get closer relations with Arab states and also changing the call to prayer from Turkish back to its original Arabic. The military accused him of planning terrorism to remain in office (ridiculous as he was popular). World leaders and the Democratic Party presidential candidate, John F, Kennedy, called for clemency, On May 31, 1961, Turkey's military hung President Adnan Menderes. .

In 2002, Tayyip Erdogan came to power and wrested power from the military. Fethullah Gulen who had established private prep schools, joined Erdogan's Development and Freedom Party and got the majority of seats in the Parliament. Turks from the rural areas flocked to the cities for jobs. Roads improved and Turkey's cinema and television replaced foreign films and TV.. The Lira dropped in value till it took thousands to buy anything. On January 6, 2006, six zeros were dropped and the new Turkish lira (YTL) was created.

The Gaza Freedom Flotilla of nine ships headed for Gaza with clothing, medical supplies and food for Gazans. On May 31,, 2011, Israeli commandos dropped by rope from helicopters onto the ships in International Waters and forensic evidence showed that the nine men killed *were shot in the back*

In 2008, three young American Jews from Berkley, California were *jogging* on the border of Iran-Iraq and arrested by Iranian border guards. One had been working in Damascus on Syrian relief and he was fluent in Arabic. It was from this area that ISIS started its push across northern Iraq and northern Syria that connected with the rebel movement the CIA abd US State Department was arming and funding to fight Bashar Assad and ISIS volunteers, weapons and money poured from Bilal Erdogam in Turkey to ISIS in Syria.. .

Twenty-nine years old American journalist, Serena Shims, mother of two little girls, saw NGO trucks carrying ISIS volunteers frm Turkey into Syria. A huge Mercedes Benz cement truck crossed over the road from the other side and hit her small car head –on cruching the front end down. . She was alive and taken to the hospital where she arrived dead of *heart attack*. She had been threatened by Erdogan's intelligence to keep silent but she released the information to Iranian television.

Vladimir Putin's Gazprom won the right's to exploit Israel's off-shore gas field that Hillary Clinton helped along from State Department by trying to destroy Syria's Bashar Assad. The US Ambassador went through Syria dispensing money for demonstrations and Bashar Assad isolating the rebels. The US Syaye Department and the Pentagon used Turkey to supply the rebels who also cooperated with ISIS.

Israel wants to divide Syria into three nationas with the Ala Wite used for the gas pipeline from Israel to Turkey's gas pipeline carrying Russian gas into Europe. By destroying Syria Israel would end Syria's right to recover Golan Heights that Israel seized in April 1967 and illegally annexed. Israel has bombed parts of Damascus in support of the rebels there.

Putin's Gazprom hold 90 % of the rights to exploit Israel's gas field, Rothschild's Total holds the other 10% The CEO of Total was in Moscow to arrange for use of Total's ships but he died in a plane crash on take-off from Moscow Airport. .

Vladimir Putin based top Russian aircraft in Syria to bomb ISIS and also attack Syrian rebels who were cooperating with ISIS. Russian planes destroyed nearly 2,000 ISIS oil tankers carrying the stolen Iraq and Syrian oil.

A Russian SU 24 may have crossed over a small piece of Turkish border at the most for 17 seconds. A Turkish F-16 fired a rocket at the Russian SU 24 flying in Syrian air space and the two pilots ailed out. One pilot was shot and killed in his parachute harness by Turkish-Syrian rebels, a war crime or murder. When a

Russian rescue helicopter came to rescue the other pilot, it was hit and destroyed by an American weapon that needed special training to use..

On July 15, 2016, Erdogan and his family were staying in a luxury hotel in Marmaris, where theauthor lived for three years, when Turkish commandos dropped by rope from helicopter to capture him. The Marmaris police fought them off.      CNN got a call though to Istanbul and a flight for Erdogan to Istanbul was arranger and phone calls that had thousands heading for bridges that were blocked by tanks. The coup was over. 191 civilians were killed and 90 military.

Erdogan ordered 40,000 detained. The leaders surrendered and Erdogan wanted mass executions without trials. There were beatings, rape, and torture on those that surrendered.. NATO, Putin, and others told him to not to execute anyone but Erdogan did and asked the parliament to make Constitutional changes giving more power. He thought he'd be supported by 50 percent of Turks that voted for his AK Party and 30% of the other party in coalition with AK Party to make 80% support him. But it was fifty-fifty with a small percentage of the vote for changes he wanted.

What is the exact number of the 6,900 military and police coup leaders has he killed? He claimed his former friend, Fethullah Gulen was behind the coup and asked the US to extradite Gulen to Turkey. The US asked for evidence of Gulen's involvement but could not present any evidence as there is none. The honest immigration police tracked ISIS volunteers crossing Turkey to the Syrian frontier, and founded the *Peace in Turkey, Peace in the World* movement with members of the Turkish Army to oust Erdogan and his corrupt ministers and clean up Turkey's corruption..

With all the detained, teachers, judges, lawyers, and all fired, and schools closed, families left without income, the Turkish economy dropped, stock market droped,  and so did the lira. Putin came to the rescue by importing Turkey's low priced agricultural food and 3.75 million Russians take their vacations in Turkey to spend billions. Turkey's strategic location was a keystone in NATO that was created to halt an invasion of Europe after WW II. Putin backed the Russian rebels who took the Donbas region with its coal to separate from Ukraine. So the US took away MasterCard and Visa making international payments difficult. The Euro has its SWIFT code to do those payments in the Euro zone and Putin has worked diligently with Brazil, India, China, and South Africa (Bricks) to create a new financial system. Putin has created a new credit card, the Mir, and wants Erdogan to accept it for use with Russian tourists and exports to Russia which have climbed to highest levels ever for exports.

Erdogan has tortured hundreds for a confession that Gulen was behind the coup but all have died saying it was the dealing of Erdogan with ISIS and corruption in his AK Party, and his son that caused them to act. Marmaris, was a main entry port for ISIS volunteers and they would be transported to the Turkish-Syrian border where Serena Sims saw NGO trucks carrying them across to border to join ISIS.

President Vladimir Putin immediately praised Erdogan and Turkey after the failed coup for saving its democracy while Erodgan has attacked the US dollar, US Government, NATO, Germany, and Holland but remains in NATO and so NATO is silent about his crime. Erdogan's move to make business with Serbia, whose history is entwined in fighting Turkey. Putin keeps silent rather than publically expressing any disagreement.

What washiddenin the Bosnian War was that Bosnia had oil which the Russians would have developed for the Serbs if they had defeated Bosnia. Erdogan is offering to develop that oil in the territory Bill Clinton and Strobe Talbot allowed the Serbs to take from Bosnia. Serbian textile mills and in privatization of spas in which Turkish investors have expressed interest and a stem pipeline to ho run through Serbi.      The meeting's agenda was discussed on Monday by Vucic and Undersecretary of the Turkish National Intelligence Organization, Hakan Fidan will discu the security challenges faced by both countries.  Serbian President Vucic and Erdoğan already met in Istanbul in early July where Vucic was one of the keynote speakers at the 22ⁿᵈ Oil Conference. ucic said that "Serbia respects Turkey's power, as it respects itself."

The possibility of another coup attempt is growing and Turkish, newspapers, journalists, and publishers are being watched by Erdogan's secret police, and there is fear

## Overview

People in the news media were called to meeting in the Israeli Embassy in Washington D.C. a day before 9/11.  They were told about 9/11 to come and to hammer on Islam as a religion of hate and that Arabs have persecuted Jews for hundreds of years.  Not true as Jews have lived for centuries among Arabs and Muslims and prospered.

In 1832, Nathan Rothschild, who is an Askenazi Jew, with no ancestral ties to Palestine, tried buying Palestine from the Sultan of Turkey.  The Sultan refused and said, "Palestine is a *Holy Land for Muslims, Christians, and Jews and Turkey would keep it way.*"

Benjamin Freedman was a high Jew and the liaison for Henry Morgenthau Sr. who manipulated Wilson to get America into WW I.   He revealed how 156 Jews at Versailles Peace Conference in 1918, dictated the peace terms and the formation of the *League of Nations.* Britain was given the Mandate of Palestine and set up the *Homeland for the Jews.*  Peace became impossible with the Jewish terrorist gangs, Lehi, Betar, Stern

175

Gang, Irgun, and Haganah killing British soldiers, Palestinians, Lord Moyne of Britain, and Count Folke Bernadotte, cousin to King Gustav of Sweden.

Bernadotte was in chargel of all the International Red Cross that inspected of all German prisons and concentration camps run by Nazi Germany from 1933 to the end of WW II. The Red Cross could enter and talk to any inmate in the prisons, prisoner of war camps, and concentration camps of Germany at any time as well as . There is not one word about poison gas used to kill Jews, Gypsies, or Greek peasants or of ovens that cremated six million Jews labeled as *Extermination Camps by Holocaust* defenders). The files of the Red Cross were rifled and records disappeared except for one document stating only 71,000 died in all the concentration camps run by Germany during WW II and it ws stated that most died from typhus.

A *you tube* video made by Jack Cole, an American Jew, who used his own money and went to Auschwitz to make a documentary video to show that the Holocaust is a money making tourist attraction for Poland with 500,000 people visiting annually. Cole tells that the bomb shelter was converted into the gas chamber *after* WW II and that tourists are not shown the hospital that looked after the health of people in this huge camp of 6,000 and who worked in the factory producing rubber tires for the German Army.

Israel has annexed at least 65% of what the UN designated as Gaza. Right after 9/11, Israel attacked its civilian population with military weapons, which is a war crime. Israel prevents Gaza from developing that rich gas field by keeping its navy on top of that rich gas field.

After BP's discovery, Israel pushed for exploration in its waters and found a gas field and adjacent to Gaza's.

Vladimir Putin has said that President Obama has financed, trained, and supplied the Syrian rebels fighting Bashar Assad.  This ISIS/daesh has hired them with money it made with of sales of oil.  Mr. Putin said he knew who was buying the oil and how much ISIS is paying each of these *rebel mercenaries (itt has become known that Bilal Erdogan has been the buyer).*  Mr. Putin states unequivocally that ISIS is run by Rand Corporation working out of the US Embassy in Ankara, Turkey.  Rand Corporation was also involved in the attempt to overthrow President Franklin Roosevelt in 1934 using Marine Corps General Smedley Butler.  Butler tells how the Marines were used by the bankers to collect on loans made to republics around the Caribbean 1912 to 1914 and said he was no more than a racketeer in doing that.

Vladimir Putin has shown the lies dispensed by the Western New media has propelled him into the leadership in the Middle East replacing the influence of the US because of the lies and destruction created by the Bush Administration.  Mr. Putin has backed up his statements with facts/  His coming to aid Bashar Assad while that is indirect opposition to Secretary of State Hillary Clinton's and President Barrack Obama's helping rebels in Libya and rebels in  Syria turn into ISIS.

In 2015, 11.5 million documents detailing 214,488 offshore entities from the Panamanian law firm ot Fonsec Mossack were leaked to the Suddeutche Zeitung of Germany and published as the *Panama Papers.*  The leaker remains anonymous as he or she would be in danger.  Politicians, drug traffickers, or celebrities wishing to hide their identity from tax offices were among the names

Celebrities, British Prime Ministers, were among thenas well as Russia's Vladimir.  Were the Panama Papers leaked to discredit Vladamir Putin for exposing the American agencies responsible for the emergence of the terror groups that have destroyed the lives of many in Syria and Iraq?

Until the break-up of the Soviet Union December 25, 1990, the KGB was the top intelligence service in the world and had infiltrated MI6 and the CIA. However, inAugust 1991 there was an attempted coup against both Michael Gorbachev and Boris Yeltsin with the Communists taking over the Duma and a tank was sent to fire a shell into the Russian Parliament building that ended the attempt to overthrow Yeltsin.

Edward Snowden has said Israel and Britain devised *Plan Britannique* to create a threat on Israel's borders. Israeli agents put out the idea that Syria should be cut up into three countries, with the coastal nation, Al Alawite, to be used for the gas pipeline from Lebanon to Turkey's pipeline that carries gas from the Central Asian republics to markets in Europe.

Mr. Obama has been duped by the CIA, Pentagon, FBI, and State Department employees. Mr. Putin gave Russian anti-aircraft missiles to the Russian rebels in the Donbas to protect them against Ukraine fighter jets. The Russian rebels used these missiles in an attempt todown Mr. Putin's plane returning from Brazil where he was in talks about BRICKS and took in a FIFA game. On his return to Moscow, his plane and the Malaysian Air flight 17 crossed the flight paths and Malaysian Air was shot down. The Donbas Russian rebels Russians are the descendants of millions Russians sent into the Ukraine after Stalin killed 14 million Ukrainians and 2 million Crimean Tartars in WW II. The leaders of these breakaway Russians were in the Bosnian War helping Serbs massacre Bosnians and Croats in the break-up of the Yugoslav Federation.

George H.W. Bush pushed for a *New World Order* which he said will *govern by a rule of law.* Remember Dan Qayle was Mr.

Bush's Vice President and couldn't spell *tomato* and was a *role* model for his sons, George W, Jeb, and Marvin. .

Mr. Putin's dramatic appearance in Syria to fight ISIS has China sending some is naval ships to Syria.. The Muslim world not connected to the US policies has been disappointed with the old Russian and Chinese policies in Chechnya and Xingjian but with the 50 attacks a day to be increased to 300 attacks a day on ISIS and ISIL has been welcomed by Muslims all over the world who are rejoicing that Russia and China are supporting Syria and Iran. Mr. Putin calls it a *shift* from unipolar leadership, but with the lies exposed, it may be that Russia will be looked to for leadership instead of America.

Mr. Putin gave Prime Minister Erdogan a 35% discount on gas sold to Turkey for the gas pipeline crossing Turkey to Europe. When Putin moved into Syria and he threatened to cut diplomatic relations with Turkey if Erdogan interfered with Russian operations in Syria. Bashar Assad's government was democratically elected and recognized by the UN while Mrs. Clinton and Israel supplied weapons, finance, and training to an unrecognized group of rebels as they did in Libya where stinger weapons given to anti-Gaddafi Libyans were fired at an American Chinook helicopter in Afghanistan and traced back to the Stinger missiles Hillary Clinton had given Libyan rebels.

Going back to May 14, 1950, Aydin Adnan Menderes became Prime Minister of Turkey and on February 18, 1952, Turkey became a member of NATO. In 1955, Menderes was assigned the duties of Foreign Minister and Turkey, Iraq, Iran, Pakistan, and United Kingdom formed CENTO which was known as the Bagdad Pact. In 1958, the military of Iraq overthrew the monarchy and killed the royal family and Prime Minister. Communists began searching for and beheading foreigners, hanging their naked corpses from lampposts of Bagdad and the military was afraid to stop them.

The author was on his way to Iran to work on the Iranian Sugarcane Plantation. It was fifteen miles to the Iran-Iraq border and the Iranian railway that the Germans built for Iran in 1934 was hauling thousands of American WW II trucks and tanks north to support Iran. The author worked for a British company and writes about it in his book, *The Hypocrites*.

Under Aydin Adnan Menderes liberal policies, Turkey began to progress at 9% annual GNP growth. Agriculture was mechanized, thousands of schools and mosques were reopened. But m military was under control of Israel and the military arrested him and executed him September 17, 1961

So when Mr. Erdogan came to power in 1998, he had to deal with a military had executed President Adnan Menderes in 1960 on charges of changing the Islamic call to original Arabic. He skillfully managed to get more freedom and Turkey began to develop and prosper and were happy with Erdogan.

But a Communist movement among the Kurdish people of Iraq began to influence Kurds in eastern and attacks were made on Turkish soldiers, that angered Erdogan. The Gaza Flotilla that sailed from . Kurdistan was the ancient nation of Turkey who want a country of their own but were split up between the Turkish Republic, Iraq, and Syria killed Turkish soldiers which got support from Kurds in northern Iraq and Syria. Israel used the PKK Kurds to attack Saddam Hussein in Iraq.

The Syrian elite support Bashar as they liked his father, Hafez Assad, who fought a bloody conflict with the Muslim brotherhood in western Syria. They are no longer a danger but it is Israel that is the danger and Israel lied in 1967 in taking the Golan Heights saying it was firing artillery from the top of Golan Heights. That eventually led to Black September and Hafez Assad

180

opening the borders of Syria to Palestinians fleeing Jordan when Israel pressured King Hossein to put down the PLO in Jordan. What Israel wanted with Golan Heights were the springs that supplied Damascus for over 5,000 years. Israel immediately dammed up the springs and turned the water into Israel. The American Ambassador to Syria began to consort with Syrians unhappy with Bashar Assad's need for internal security and Israel helped in the demonstrations the Bashar Assad saw as related to events in Iraq and the execution of Saddam Hussein. He gave the secret phone number to Muammar Gaddafi's phone to the Americans thinking that they would merely capture him and leave him alone.

*WikiLeaks disclosed* the US gave millions to Syrian opposition groups up to 2010. State Department has control over all overseas shipment of weapons and as Secretary of State, Hillary Clinton was heavily involved in handing weapons out that went to unorganized groups in Syria, to pave the way to create *peaceful demonstrations* in 2011.

A video on you tube showed retired General Wesley Clark telling about the plan to subvert seven nations of the Middle East: Iraq, Libya, Yemen, Somalia, Sudan, Syria. and Iran

Even though Mr. Putin has the rights to exploit the gas field off shore Israel he came to the rescue of Bashar Assad when he could have let the United States tear Syria apart and exploit the Israeli and Gaza gas fields. But he played it morally right and come to aid Bashar Assad. To mollify Mr. Netanyahu, he explained that the military equipment he sent Syria is nothing for Israel to be concerned about as Syria is more concerned about ISIS and the Syrian rebels.

Mr. Erdogan was angry with the Israel attack on the Gaza Flotilla coming from Turkey by Israeli Shayetel Commandos that killed 8 Turkish civilians and one American. Israel claimed they were attacked by men with iron bars and knives but when forensics examined the bodies, their wounds were all in the back and Mr. Erdogan expressed anger at Israel and threatened to cut off diplomatic with Israel.  Israel sent a mission to Turkey and all was smoothed over but corruption began to grow in Mr. Erdogan's government.  With the arrest of five members of his cabinet by Turkey's police, Mr. Erdogan snapped back and began shuffling the police chiefs and loyal police around who were highly moral and upholding the law and firing some.

Mr. Netanyahu and the Likud were tired of waiting and put *Plan Britannique* inaction (the *Hornet's Nest)* ISIS and ISIL.  Al Baghdadi was trained by Israel in the Koran and mind control techniques in Dimona, Israel and the ISIS spread rapidly from near the border of Iraq and Iran where three young American Jews were captured by Iranian border police when they went jogging and accidently crossed the border into Iran.  The Sultan of Oman paid $350,000 each for their release, and he has also given asylum to Ayesha Gaddafi's daughter who the Libyan rebels have requested her arrest through Interpol in Paris.

But with the northern Iraq oil from the Kurdish area and western Syria's oil fields, ISIS has been shipping that oil into Turkey which is bought by Erdogan's son, Bilal and sold to the Turkish oil company at $8 a barrel, and exported to Israel from the Turkish port near the border of Syria where two Turkish Air Force Jets shot down the Russian SU 24vand Mr. Putin was vexed to say the least.  A Russian helicopter was destroyed by a missile from Turkmen rebels in that part of northern Syria with a missile given to the Syrian rebels by the American government.  Mr. Erdogan tried invoking Article 5 of NATO agreement that said if one nation is attacked the rest will come to defend it.  NATO's response was that Russia did not attack Turkey and has since tried to get Mr. Erdogen from being too extreme in his treatment of the military who took part in the attempted coup and release all police, immigration officials, and journalists who have dared to criticize his government.  Over 100,000 have been arrested, detained, teachers fired, and workers laid off.

The purchase of ISIS stolen oil by Turkey gave ISIS (ISIL) money to buy weapons and munitions, which Mr. Putin very cleverly used Russian TV carried internationally to tell how America was supporting and financing rebel groups who were terrorist organizations.  Serena Shims, an American Lebanese reporter on the border of Turkey and Syria, was murdered by Turkish Secret Police under the control of Mr. Erdogan after her small car was in a front end collision with a cement truck.  Mr. Obama's administration has kept from the news media.

Mr. Erdogan has prevailed upon NATO countries to view Bashar Assad as a killer of his own people. Mr. Obama began training Syrian rebels and supplying anti-tank missiles with shaped thermite charges that melted the Bashar Assad's tanks. Iran has supplied the Houtis Rebels of Yemen that melted the American tanks of invading Saudis.  Some have connected the torture and kidnapping of Muslims by President George W. Bush to the massacres of Bosnians by Serbs and Boris Yeltsin's massacres of Chechens to the sudden immergence of ISIS in the northern Iraq that spread into northern Syria.  Some 7,000 Russians have joined the ISIS because of the war crimes of Boris Yeltsin that Strobe Talbot and Bill Clinton covered up.

July 13, 2015, Turkey apologized for shooting down Russian SU 24 that allegedly crossed into Turkish air space for ½ a minute.  Then came the July 15, 2015 attempted coup to overthrow Mr. Erdogan that failed.  It a group of top officers at Incerlik Air Base were the coup leaders and Incerlik is a Turkish air base it shares with the US Air Force to use in accordance with NATO operations.  Mr. Erdogan has claimed a former friend and ally, Fetullah Gulen is head of a terrorist organization he has labeled FETO, and calls upon the US Government to extradite him to Turkey for trial. The response from America has been to ask for the evidence. Mr. Putin's response was quick to proclaim Mr. Erdogan as a hero to democratic rule.

The coup was planned by faction within the Turkish military were angry over the corruption and the wars being financed with stolen ISIS oil. They called their movement *Peace at Home* and attribute it to Ataturk's statement *Peace at Hone and Peace in the World.* Ataturk escaped Istanbul to a port city on the Black Sea and formed and army that began to drive out the occupation forces that partitioned Turkey among them. Nobody wanted to go back to war again after WW I so he agreed to a peace made in Switzerland for a republic and an end to the Sultanate. All was in accordance with a Homeland for Jews as the last thing the Zionists wanted was a million man Turkish Army marching on Israel and that led to the Iran hand of the Turkish military the people of Turkey became sick of. But the sympathy for Palestinians is strong as it is with the World Peace Movement founded in 1948 and has 100 member nations today.

Mr. Erdogan tried to seize the 50 to 90 atomic warheads the Americans had stored in Incerlik and shut off electricity and water to the base. The American went alert and turned on their own emergency generators. Mr. Erdogan then turned to NATO and tried to get them to invoke Article 5, arrested, detained Turkish police, military personnel, taken away education certifications from 21,000 teachers in private schools, and turned so far against America that he is telling people to buy gold and do away with their dollars.

With tourism down, the Turkish Lira down, and the Istanbul stock exchange down, Mr. Erdogan looks forward to Mr. Trump's inauguration and the plan to put his manager of Trump properties in Turkey as the new US Ambassador to Turkey.

The author knows a lot more than he has written in his other books. There is the death of a Turkish man who was scuba diving and drowned. A Danish scuba diver was charged with murder. Well, he was also Danish Hell's Angel's member and connected with the Turkish political party that is also tagged as *Grey Wolves*. Both groups are into drug trafficking in Turkey as is the Israeli Mossad. And both groups do assassinations for the CIA assassination unit. That is all I will say on this subject.

But Turkey is a very comlex country. In WW II, thousands trucks came from Bulgaria with foods, clothing, and medicines to cross Turkey and go into Russia. No one could find out who was sending supplies to Russia from Germany but the truck brought back Russian Baku oil. Turkey also let Germany use their airfields to bomb Stalingrad and Marshall Zhukov wanted Stalin to declare war on Turkey. Stalin said he would not as much of the Russian Army depended on Muslims from central Asia to hold back the Germans.

But millions of Jews came from Europe at the end of WW II to get into boats in ports of Turkey to sail to Palestine. And those ports are now the tourist spots of Turkey that bring in billions in Euros, Pounds, and dollars and offer tax free shopping for cruise ships on the Mediterranean.

Barbara Honegger tells how Flight 77 that off from Dulles Airport with 81 passengers for Indianapolis went off the radar somewhere over the conjunction of Ohio, Kentucky, and West Virginia. She interviewed many Pentagon employees and the most impressive was April Gallop who recalled many explosions going off in the Pentagon to bring down the sections and bolster the lie that a Flight 77 was hijacked and flown into the Pentagon.

185

The FBI supplied clips from security tapes showing an explosion dated September 12, 5:30 pm.

The North Tower was brought down by thermite placed on the steel girders inside the tower's perimeters and a mini-nuke was placed on the basement floor. *No plane hit the North Tower*. The Mastermind of 911, Mossad Colonel Mike Harrari, died of cancer September 21, 2014 and Israeli Prime Minister, Ariel Sharon, went into a coma for years and died without reawakening, both behind 911 yet Muslim jihadists were

Rebecca Roth, a flight attendant for many years, wrote that the radio transmission of flight attendants Ong and Sweeny was faked as passengers would be screaming as the plane descended so rapidly from 10,000 feet to hit the North Tower and they were not doing their duties as flight attendants.

The photo of the crater below the twin towers was kept off the internet until recently. And NORAD sent out an alert about the missile traveling six feet above the ground, heading toward the Pentagon was ignored, A video showing the plane swallowed up after hitting the South Tower could not have gone through steel beams. Israel does not want an investigation of 911 as it would show Americans that Israel has taken trillions what would have paid free medical, free education, and cancel all credit card debt of Americans.

On February 5, 2015 the author went to print shop Multicity den Haag, Toussaintkade 72 and ordered an A4 size of his book, *Treachery* along with the author's request to act as the representative of the people of the United States to charge Israel invasion of American air space, commit mass murder, endanger the health of millions in the New York metropolitan area, destroy billions of dollars of property, and cover the theft of 2.3 trillion dollars by Israeli-American citizen, Rabbi Dov Zakheim, and 850 million in gold that disappeared from the bomb proof vaults under the WTC buildings.

The author included his documentary of the *Bosnian Conflict, Deadly Diplomacy,* where charges of genocide against, Serb Major Milar Aramovic, were dismissed by Judge Richard Goldstone of the IWCT.  However, Aramovic requested and *got 12 kilos of gold to do it.*  He also told a young Croat that if he didn't want to help kill 8,700 men and boys at Srebrenica, *"You can join them if you feel any sympathy for them."*  The young Croat said he killed 55 man and boys but his conscience bothered him.

The author included a print-out of his American heritage, *The Love of a Father,* to show he is an American and could represent America in charging Israel of the crime of 9/11.
The package was address to The Honorable Patrick Robinson of the International Court of Justice with a letter requesting he be allowed to represent the United States in charging Israel with attacking a member nation of the United Nations.

The author wanted UPS or DHL to deliver the package but the head of Multicopy den Haag said they would deliver the package to Jurist Patrick Robinson and the author could pick up the receipt the next day, but on Saturday, Multicopy was closed.

So the author traveled to Berlin where he had another package with the same material created and sent to The Honorable Patrick Robinson of the International Court of Justice by Deutsche Post with the facts and evidence that 9/11 was a False Flag operation perpetrated by Israel.

The International Court for Justice in The Hague deals only with crimes committed by one member nation of the United Nations and not individuals. The International Court of Justice has jurisdiction over both Israel and the US as it was founded in 1946 as part of the UN when it was established.  The author stated he
187

was pressing charges against Israel for the people of the United States as the authorities were ignoring the facts and have committed crimes of kidnapping, torture, mistreatment, mutilation, taking of hostages, outrages on personal dignity, and executions without due process. The purpose of detaining and prosecuting those in Guantanamo and prisons outside America was for the purpose of denying them the rights they should have under America's Constitution.

Years before 9/11, in June 1989, the Prime Minister of Trinidad and Tobago, Mr. A.N.R. Robinson, revived the idea of permanent courts to try individuals of a country with crimes against humanity, torture, murder, slavery, rape, and sexual slavery, and the UN General Assembly began work on a draft and the UN Security Council established two *ad hoc tribunals* in the early 1990s.

President George W. Bush, an accused war criminal himself, got the IWCT to give immunity for all Americans accused of war crimes and immunity for foreign nationals working with American companies working in Iraq. The US is a signatory to the IWTC but Congress has not ratified its membership. IWCT can only prosecutes individuals who are serving a country that is a signatory to the IWCT which is why mercenaries are considered as combatants but not under any jurisdiction and Serbia, Russia, and the US use mercenaries.

In WW II, General Henry Arnold had the idea of using war weary bombers converted into drones to bomb German submarine bases and V1 rocket bomb bases covered with steel reinforced concrete. General Doolittle ordered Flying Fortresses (B17 bombers) stripped of all armaments and filled with 15 tons of explosives, Electronic equipment was installed that would allow the drones to be controlled by a mother ship flying nearby. The drones were to be flown into the air by a pilot and co-pilot who would then parachute from the drone and its controls would be taken over electronically by the mother ship and guided into re-enforced steel in concrete bunkers covering German submarine bases. That 15 tons of explosives never made a dent in the German steel reinforced concrete. President John F. Kennedy's elder brother, Joe Kennedy Jr. died in that operation when his plane exploded before he could parachute from the plane.

John Lear says that it is impossible for anyone who had only learned how to fly one engine propeller driven Cessna to take over a multi-engine jet passenger plane and know how to turn off transponders and black boxes and know how to read all the instruments and which switches to flip.

Additionally, the aluminum fuselages of the planes could not have brought down in a free fall the twin towers 90,000 tons of steel encased in a steel reinforced concrete shell. In the center of each tower where 169 high speed elevators took 10,000 workers to their offices every day that were surrounded by steel shafts resting on bedrock

An Israeli company held the contract for Logan Airport security and Logan Airport where the Mossad got the Arab *students* past immigration most likely because an Israeli company runs security as one was previously denied entry into the United States. American Airlines flight from Amsterdam's Schiphol airport with British shoe bomber Richard Reid was diverted to Logan Airport and escorted by two Air Force jets. Same with the underwear bomber who flew from Schiphol. Schiphol Airport is where Mossad offices are in Holland. Boston is where the Boston Marathon was bombed by two brothers from Kirghizstan. Movie producer Nathan Folk and others have said was a False Flag. One brother was killed by the police in a shoot-out and the younger brother was convicted and sentenced to death for the bombing. His appeal is pending.

Flight 77 had 64 passengers on board (including 5 hijackers and a crew of 6) and 125 people in the Pentagon were killed by Hani Hanjour, a trained pilot, alleged to have assumed control of the plane which is an absurdity.

A Wikipedia reported dozens of people witnessed the crash and that news sources began reporting on the incident within minutes. The Wikepedia author of that piece is way off.

Thirty-three minutes after the missile hit at 10:10 a.m., a section of the Pentagon collapsed. A firefighter is pictured by a big round hole that swallowed a jumbo jet three stories high. At the bottom of this hole to the right is a wing of *the supposed plane. April Gallop* carried her baby through that hole and flight 77 from Dulles was tracked by radar to southeast corner of Ohio while the two flights from Logan landed in Cleveland, Ohio on September 11, 2001.

**Thirty three minutes after the crash** (10:10 am) three sections of the Pentagon collapsed and the next day, photos taken from the roof of the Pentagon showed luggage and clothing on top of the debris. The damaged section of the Pentagon was rebuilt in 2002, and reopened August 15, 2002.

A photo taken from the South Tower of the hole in the North Towers shows a girl waving *from the hole made by an airplane.* How did she escape from being killed by an airplane that created that hole? The NIST report does not include Jennifer Oberstein's call to Katie Couric or mention the girl waving from the hole.

The *plane* that hit the Pentagon was a missile and a clip of the missile's exhaust came from a security tape gathered immediately by the FBI. Solicitor General of the United States, Ted Olsen, claims he received a telephone call from his wife,

190

Barbara Olsen, who was on flight 77. The FBI said no call was made and the airline says that seat phones are on their planes. Ted Olsen has made three different explanations and Barbara said the hijackers used box cutters (used to open cardboard boxes and knives to force all passengers and crew to the back of the plane. But Barbara Honeggar said that radar tracked Flight 77 to SE Ohio. Ted Olsen married *Lady Booth* in 2002, who is a lady of mystery as nothing is known about her.

Edward Snowden has given information from NSA files that information taken from phone calls made by Palestinian-Americans in America were passed on to Israel's 8400 unit to blackmail or pressure them to spy for Israel. Snowden reveals on page 144 that the *Hornet's Nest* was an Anglo-Israeli plan to concentrate *extremists* and fake Caliph in the Middle East.

A video on you tube has Secretary of Defense Donald Rumsfeld revealing that 2.3 trillion dollars was missing from the three trillion dollars given to Pentagon Comptroller, Rabbi Dov Zakheim on May 4, 2001. A Russian missile hit the steel reinforced concrete office wall behind which 29 accountants were killed and the Pentagon's accounting books were destroyed as were the back-up files in Building 7 on 9/11.

The number of murders committed by returning soldiers at Ft. Carson shows the psychological damage done to men who have been trained to kill, The Hollywood movie, *Sniper,* was based on the true to life experience of Chris Kyle who was a sniper in Iraq. Chis was trying to help Eddie Ray Routh, a veteran traumatized by combat experience when Ray turned on Chris and killed him. The jury found Eddie Ray Routh guilty of murder. He pleaded insanity but the jury rejected his plea.

Frank Sturgis was 19 years old when he joined the Marines in 1942 and was trained as a sniper to be dropped by parachute behind Japanese lines where he targeted Japanese

191

officers.  After Sturgis was discharged from the Marines, he was hired by Jewish terrorist groups in Palestine as a sniper.  After the Mossad was formed in 1948, Sturgis worked with them.  He was the sniper behind the wooden fence at the top of the grassy knoll October 22, 1963.  Sturgis also trained Fidel Castro's original group with CIA approval.

The US Department of Veterans Affairs states that on an average day 22 veterans commit suicide as a result of PTSD (Posttraumatic Stress Disorder).  Combine this with those who murder shows training people to kill does have an effect.

The Oklahoma City bombing was to blame on Muslims and stop the outpouring of sympathy for Bosnian Muslims being slaughtered, raped, and robbed. President Clinton did nothing because his Under Secretary of State Strobe Talbot said supporting Boris Yeltsin as first president of Russia was more important than stopping the murder of thousands of Bosnians (1991-96) which led Osama bin Laden to use his money to send the CIA trained Mujahedeen to Bosnia and which were clearing Serbs out of Bosnia which the CIA then gave them the name Al Qaeda as a *terrorist group.*

Slogans are the way to start wars and the slogan for WW I was the *War to End All Wars.*  Russia, Great Britain, France, Germany, and Austria were the only great powers and the plan was destroy Germany and Austria, get Turkey in the war to get Palestine as a *Homeland for the Jews* and the other three great powers will be indemnified by Germany for instigating the war, planned in London with the signatories of the Secret Treaty.  The murder of Archduke Ferdinand in Sarajevo was to get Turkey into the war to defend the Bosnian Muslims from the Serbs.  But the Sultan refused and said he could not dishonor Turkey's *British Protectorate* given to Turkey in 1878.

To get Turkey into the war, German Banker, Paul Warburg, goes to Kaiser Wilhelm and persuades him to send two German cruisers to Turkey as *gifts from the German people to the people of Turkey.* After Turkish flags are raised and Germen crews are wearing Red Fezzes, the German captains sail for Russian Black Sea ports, bombard them and Britain, France, and Russia declare war on Turkey. Paul Warburg wrote the Federal Reserve Act and Warburg Bank is one of nine Jewish Banks of the Federal Reserve which is financing the sale of American weapons to the countries fighting Germany and Austria.

Paul Warburg go to Wilhelm to say Turkey needs to be brought into the war as it can go over the Caucuses and get the oil that the Kaiser needs for his German navy. He says the Turks ordered to battleships from Britain which they paid for but the British would not deliver them because they said Turkey might use them against Britain. The plan is to send two new German cruisers Istanbul, fight through the blockade at the Dardanelles, and in Istanbul say they cannot go back and the Kaiser is giving the cruisers from the German people to the Turkish people and celebrations are all over Istanbul. Turkish flags are raised, the German crews given red fezzes, and when the 20 Turkish officers are on board, the German captains sail for Russian Black Sea ports, bombards them and Russia, Britain, and France declare war on Germany.

Winston Churchill is told to plan the invasion and the plan is a diversion invasion at Gallipolis with Australian, Indian, and New Zealand troops. The bombardment has the Turkish troops retreating. But on meeting Colonel Ataturk, he orders them to turn back and they storm the British lines with human wave tactics that cost Turkey over 300,000 men. The British are driven back to the beaches and machine guns from Germany are brought in and the British lose over three hundred thousand men.

But the Russians come of the Caucuses and occupy eastern Turkey.  Bringing Turkey into the war has not helped Germany.

Warburg goes again to *Kaiser Wilhelm* to tell him he can get 600,000 Jewish conscripts *to desert the Czar's Army.*  Small crystal radios inform the Germans where the desertions occur and six million Russian soldiers are lost.  Czar Nicholas abdicates and Prince Kerensky takes over but continues the war.

So now Warburg tells Kaiser Wilhelm to get Lenin into Russia to topple Kerensky and Lenin is brought by boat from Stockholm to Helsinki and by train to St. Petersburg.  Lenin makes speeches but nobody pays any attention to him.

Jacob Schiff of the Kuhn Loeb bank, also of the Federal Reserve, calls Trotsky from New York to Washington, and gives him twenty million in gold to buy demonstrations for Lenin. America is brought into the war to save Britain and Balfour sends a signed copy of his Balfour Declaration to Lord Rothschild that is called the *title to Palestine.*  The Federal Reserve made of nine Jewish banks of which the Rothschild Bank had established two previous central banks for the United States in 1791 and 1816 that were called the Bank of the United States and the Rothschild are behind the Federal Reserve Bank that is financing the sales of weapons from the War Industries Board managed by Bernard Baruch and Jacob Schiff and Paul Warburg of the Federal Reserve are undermining Germany and Russia preparing Europe for Communism.  When Communists in Germany began a fight in Germany to establish a republic, Warburg tells Wilhelm to abdicate and run to neutral Holland for asylum because if the Communists win, they will kill him and all his family like they did his cousin, Nicholas Romanov and his family after Nicholas had abdicated but Lenin still had him killed.  The Rothschild will get Palestine free and legal through the League of Nations, whereas he had offered money to the Sultan in 1832 and asked President

Benjamin Harrison to put pressure on the Sultan to sell Palestine. In addition, Rothschild got all the rice farms of America by selling the bumper rice crop shipped out of India in 1914 for three years with rice farm owners walking away from their farms and Rothschild got another collective farm in America.

America is brought into the war to save Britain and Balfour gives a signed copy of his Balfour Declaration to Lord Rothschild that is called the *title to Palestine*. More than 88 million lives were lost in WW I.

A British general says he can crack the German line with the same tactics and is given 500,000 Indian soldiers. They are lined up and given a rifle and told to advance to the German lines and if they turn back, they will be shot. In 14 days all 500,000 are dead. The German line was invincible.

At the end of the war, 200,000 Indian men out of 950,000 men were returned to India where fifty million men, women, and children were starved to death in 1914 as no one shipped the Indian rice back into India. So now you know why hundreds of million Indians looked for Hitler to free them from the British.

The man who managed his successful presidential campaign was Col. Edward Mansell House of Houston, Texas. House was the manager of the corporate farm owned by the International bankers that was created by the B'nai B'rith Carpetbaggers for Nathan Rothschild after the Civil War for pennies on the acre. Edward Mandel House handled the cotton sales from that collective Rothschild farm and was head of the Texas Rangers by Governor Hogg he got in the Governor's mansion in Austin, Texas, and House could get any man running for a political office elected.

Samuel Untermeyer paid the $40,000 for the breech of promise sued filed on him by the wife of a former neighbor of Wilson who divorced her husband to marry President Wilson. But when Wilson didn't, she sued *and* Untermeyer paid in return for a *favor.* The favor was to appoint Louis Brandeis as the first Jew to serve on the Supreme Court.

Benjamin Freedman was liaison for Henry Morgenthau Sr. and tells how the Jews surrounding President Wilson got America into WW I. Mr. Freedman was a Jew and proud of it. He tells how Samuel Untermeyer called for an International Congress of Jews to meet in the Bilderberger Hotel near Amsterdam to declare war on Germany in June 1933. Franklin Delano Roosevelt began preparing America for war to save Britain as Britain had the Mandate of Palestine as a *Homeland* for the Jews. If Germany defeated Britain there would be no Israel as the Nazi Party in 1944 had the elimination of Empires in its plank and all colonies would be independent and free too. Benjamin Freedman was an American patriot.

While the B'nai B'rith is a Jewish Community service organization, it was also the Confederate spy ring. The Civil War ended when the B'nai B'rith and Communists in the Republican Party were told to end the war.

Simon Peres in April 1995 revealed to David Frost on BBC TV Communism was created by the Khazar Jews of Russia because of the Programs. Benjamin Freedman said 98% of the Communists in Europe were Khazars and not Jews.

Wayne Madsen is a Washington Insider and produced a video on you tube *Voices from the Grave.* He tells about Deborah Jeanne Palfrey who created the Pamela Martin and Associates Escort Service. Her girls were picked up limousines that took them, to rendezvous with Washington Power Brokers, officials in the CIA and government offices. Deborah called her girls *gals*. They were to charge $300 dollars plus tips for 90 minutes. Deborah stipulated they were not to drink or take drugs during their appointments, wear neat pants suits, sensible heels, discreet jewelry, be on time, and have that *Ann Taylor look*. Her girls ranged from a college professor, Naval Academy Instructor, a legal secretary, and a financial consultant (whose pseudonym was Rosslyn and looked like a brunette Farrah Fawcett). Her girls had to be over 23 years of age, have a college degree, a day job, and not give any clue as to their real name or their day job. They were to send her half their fee of $300 plus tips by Postal Money Orders under $800 to her home in California.

Deborah could not understand why the government singled her out, so she submitted a file listing 83 brothels and escort service that she knew of in Washington to the court on September 10, 2007. There were many phone calls from the McLean, and Langley area of Virginia in her 46 pounds of phone records the FBI overlooked when they took her computer.

Troy Burris (IRS) and Maria Cotillion (postal inspector) began investigating Palfrey in June 2004. She was cooperating with the government but they didn't take her phone records so she offered them for sale. Judge Judy Kessler issued a protective order blocking the sale of her phone records which was overturned.

Deborah then gave her records to Dan Moldea of Hustler Magazine USA and he found Congressman Bob Livingston's name (he was about to become Speaker of the House) and Senator David Vitter called for her service five times when he was a Representative before he became the senior senator from Louisiana. He ran for governor of Louisiana after Bobby Jindal's term was over but he lost to the Democratic Party candidate. Both Jindal and Vitter received Rhodes Scholarship like Bill Clinton, Strobe Talbot, and Dov Zakheim to study at the University of Oxford in 1972. Dov continued on in London to study at the London School of Jewish studies, 1973 to 1975.

Dr. Harlan Ullman was advisor to the Pentagon's Business Transformation Agency. Deborah called him *a most disagreeable man.* Ullman coined the phrase "shock and awe" to describe the bombing of Baghdad with 400 missiles encased in depleted uranium steel alloy so tough it held the explosive charge inside until temperatures reached nuclear fission temperatures. Ullman showed no remorse of guilt with his boast that there'd be no safe place in Baghdad for women, children, and civilians with 400 cruise missiles raining down on Baghdad for 24 hours.

Deborah argued she had no control over escort's sexual conduct but was convicted of prostitution and awaiting sentencing when she went to Florida to see her 78 year-old mother. Deborah never made it to the front door. Her mother found her body hanging from a bar in a shed on her property.

Alex Jones on his radio talk show asked if she was planning to commit suicide. Deborah answered emphatically "NO" three times. She was convinced that she would win an acquittal.

The scandal of Deborah Palfrey's *Pamela Martin Escort Service* was reminiscent of Madame Claude's service in France. Her clients ranged from government officials to wealthy Middle Eastern leaders and businessmen. When the scandal broke in France, it brought down the government of Valery Giscard d'Estange. The inclusion of Deborah Palfrey's escort service in this book is to show the risk of being exposed could have many collaborating CIA and fbi murders and Israel's 9/11 Black Ops. How much did the FBI know about prostitution in Washington? The CIA and Israel use *honey traps* to trap, get intelligence, or collaboration or cooperation.

A sham charge of rape by two Swedish women was used against Julien Paul Assange when it was conceptual. Advised he could be extradited to the US if he went to Sweden to answer the charge, he took refuge in the Ecuadoran Embassy in London where he has been cooped up for more than five years as a guest of the Embassy. The Embassy has been surrounded by London

Police all this time but an attempt to assassinate Assange was thwarted by followers who saw two men sneaking up the back stairs of the building the embassy is housed in. It has turned Assange into a personal foe of Hillary Clinton and President Barack Obama.

The author wrote in his semi-biographical work about Honduras, *Manipulation and Deception*, of being offered any job in the world that he could have any job in the world he wanted if he joined this powerful secret society that is inside the Mossad. That's a pretty heavy offer and they knew the author was capable of handling a top position as he had written good studies on economic studies used by the World Bank and other international companies.. But for that offer, it would first have to go through Hossad Col. Mike Harrari who has admitted he planned 9/11 and he had the ability and resources to do it.

How powerful was Mike Harrari?


The author had a friend who got an even bigger offer and has a former classmate who told of meetings where the reduction of the world's population by one third or two thirds was essential as the present size of the world's population made it impossible to completely control.

The repercussions of 9/11 as a False Flag Operation have not ended. Former President Bush and Vice President Dick Cheney have no shame about permitting and encouraging torture or detainees who they knew were innocent. No problem in kidnapping men and holding them indefinitely. No problem in inviting Prime Minster Netanyahu to speak before Congress on the importance of armed conflict with Iran.

Russian nuclear intelligence analyst, Dimitri Khalezov, has said that UN inspection of Iran's nuclear energy program can prevent Iran from making a nuclear weapon and Israel knows it.

President Bibi Netanyahu has used the ignorance of politicians and mid-level security people to support Bibi's assertion that Iran is aiming to make atomic weapons. Bibi has no proof and in his assertions he harangues on anti-Islamic rhetoric, but Israel bombs Syria in support of the Syrian Rebels and supplies ISIS and treats its wounded in Israeli hospitals.

International inspection and control over all nuclear technology was what President Kennedy proposed with his nuclear non-proliferation treaty Israeli Prime Minister David ben Gurion refused to sign and then gave the order to have President Kennedy killed. All Middle East nations have declared a non-nuclear weapons zone. After President Obama and Secretary of State, ohn Kerry got Iran to agree to inspection and control, Mr. Obama asked Israel to agree international inspection and control. Israel used nuclear weapons on Iraq and the US has used depleted uranium alloyed steel in Bosnia resulting in an increase in cancer among Serb villages on which it was used. And birth defects from Israel's atomic missiles fired into Iraq have been documented.

Israel has used military weapons on civilians in Gaza, including white phosphorus. Israel was pushing the US military to grab Pakistan's nuclear weapons facility. The Hillary Clinton and Obama support of the Libyan and Syrian rebels has left both countries unstable with refugees flooding Europe. Europe calls them *migrants* and not *refugees* which is how they can shut their borders to them.

In the world of intelligence agencies there are Trojans, decoys, scapegoats, and expendables. On the internet there are two stories of who El Baghdadi is. The first story is that he is a Jew by the name of Eliot Samuels or Samuel Eliot. The other story that El Baghdadi was a detainee and tortured at Camp

Bucca by the US. The Mossad acted as advisors on the torture and that is how the Mossad recruited El Baghdadi.

The author is well aware of the persecution of Muslims in Britain, France, Australia, and the United States where many Muslims have immigrated. Black Ops targeting Muslims were begun during the Bosnian War as Israeli Lobby was shocked by outpouring of sympathy for Bosnians and the campaign to get Clinton to lift the arms embargo placed on Bosnia. The embargo was illegal by the by-laws of the United Nation and it was President George H. W. Bush's Secretary of State, James Baker, who he offered to help Milosevic to hold Yugoslavia together. The campaign to lift the arms embargo on Bosnia had the Israeli Lobby working to unseat Congressmen and Senators sympathetic to Bosnia. As a result no Congressman or Senator today will listen to anyone beyond his constituency.

Jewish Organizations in America have charities, civil rights groups, human rights groups, and holding high offices in Freemasonry. The B'nai B'rith claims the KKK is anti-Semitic but Jews have always been in the KKK. The B'nai B'rith used the KKK to create racism and institute tenant farming to work the huge collective of the Rothschild's in the South that Edward Mandel House managed its cotton sales.

Edward Snowden worked for Booze, Allen and Hamilton copied records of the NSA's *Prism* program. Dov Zakheim resigned his comptroller of the Pentagon when the investigation on the 2.3 trillion dollars got too hot. Booze, Allen, and Hamilton hired Dov Zakhein, and in 2008, President George W. Bush appointed Dov Zakheim to the Commission on Wartime Contracting in Iraq and Afghanistan. In 2010, Zakheim retired as a *Senior Vice President* of Booz Allen, Hamilton. It was a standard practice by Booze, Allen, and Hamilton to hire retired

Pentagon and military officers after President Kennedy's assassination which Congress has recently condemned.

Barbara Honegger investigated the United Flight 11 that hitthe Pentagon and says if any of those involved in that flight have a change of heart and talk, they will be killed and their families reduced to poverty. She traced PNAC roots to Leo Strauss of University of Chicago in 1962. Many of his students were involved in founding PNAC (Project for New American Century) to keep America perpetually at war.

Professor Dr. Daniele Ganser worked in research on matters of economic and strategic importance behind wars of the 20th Century. He reveals that immediately after the 9/11 attack, NATO declared Article Five was in effect. Article Five states that if any member state is attacked all are attacked and be on the alert. NATO is the world's largest military force with 28 member nations around the world, not all are located in the North Atlantic. NATO's fighting forces have gone into Afghanistan. There are 138 nations in the world whose combined populations recognized Palestine while NATO nations have refused to recognize Palestine. Israel's attempt to move the US 6th Fleet, attached to NATO, and based in Naples moved to Israel erupted a firestorm in Italy.

NATO was created in 1948 to defend Europe against the Soviet Union in 1949 developed into *Operation Gladio* was a secret army that was left behind in event Russia overran Western Europe. *Gladio* operated in Western Europe, non-members Sweden and Switzerland and was in contact with Finnish intelligence.

Dr. Daniele Ganser, teaches History and the Future of Energy Systems at the University of St. Galien, wrote a book about *NATO's Secret Armies.*in 2004 exposing *Operation Gladio* that made False Flag operations in Western Europe to blame on Communists.

Dr. Ganser's research on how energy sources are used to manipulate countries into wars and that led to the discovery of why Japan attacked Pearl Harbor on December 7, 1941. It was President Franklin D. Roosevelt who wanted to get into the war in Europe and used the German-Japanese defense pact to push the anti-war sentiment in Congress to do it. FDR confided with his Chiefs of Staff, Don Smith who kept it secret for nearly all his life how Roosevelt pushed Japan into attacking the US by putting a trade embargo on Japan that cut off its supplies of oil from the US. Japan needed that oil for its airplanes, submarines, tanks and vehicles and America was the biggest exporter of oil before WW II. Without oil, Japan would lose its war with China that was started over soldiers dressed in Nationalist Chinese uniforms firing on a Japanese soldiers at a rail line in Manchuria. But could they have been Chinese Communists dressed in Nationalist uniforms.

After France surrendered in June 1940, Germany turned France's colony of Indochina over to the Japanese to manage but when the Netherlands fell, Queen Wilhelmina fled to England and established a Netherlands government in exile and the Dutch navy there was turned over to the Dutch East Indies. After Roosevelt put an embargo on oil exports to Japan and Japanese needed that oil of it would lose its war with China. Roosevelt sent a harsh letter to the Emperor of Japan warning him that if Japan went into Indonesia America would oppose the Japanese with force and the British had broken the Japanese code and so Roosevelt, who had pushed the Japanese to the wall knew the day and hour the Japanese were going to attack Pearl Harbor.

Before Pearl Harbor diplomatic protocol required all nations to declare war before going to war. On the morning of December

203

7, 1941 the Japanese diplomats assigned to Washington, went to the White House to deliver a Declaration of War. The British had broken the Japanese code and delivered the information to Roosevelt that the Japanese were going to attack that morning. Roosevelt kept them waiting until he heard by telephone that the Japanese had attacked Pearl Harbor and then met with the Japanese diplomats, denounced them for perfidy and went on radio to tell America and the world that Japan had *stabbed America in the back*. President Franklin Delano Roosevelt deliberately sacrificed thousands of lives to get America into WW II and he destroyed historical diplomatic protocol so that a sneak attack by another country is expected all the time everywhere in the world.

Benjamin Freidman was the intermediary for the Jews under Henry Morgenthau Sr. that surrounded President Wilson. Bernard Baruch was born South Carolina and the family moved to New York City where he was educated. Financier Bernard Baruch became a legend for making a fortune speculating in the stock market and became the dominant financial supporter of the Democratic Party that voted as a block in the South. Baruch saw *bad boy Woodrow Wilson's* predilection for women (now becoming an issue in Congress for male senators and congressmen being accuse of sexual harassment and abuse, and got Professor Woodrow Wilson into the presidency of Princeton. From Princeton to Governor of New Jersey, and then kingmaker, Colonel Edward Mandel House from Houston to run his campaign. Mandel House was the manager of the huge *Corporate Farm the Rothschild created from plantations the Carpetbaggers created with Rothschild money John Wilkes Booth brought down from Quebec to avoid US Customs, and of course Booth was manipulated into killing Lincoln to keep the South poor as Lincoln's plan to rebuild the South died when he was killed.*

204

J.P. Morgan was used to finance and encourage ex-president Theodore Roosevelt to form the Bull Moose Party and draw away votes from President Taft and Wilson won in a 3-way vote.

In June 1916, the German submarine blockade of British ports cut off petroleum and food imports has Britain ready to surrender.  Kaiser Wilhelm sends his demands for the surrender. All nations stop fighting and return to their original boundaries. There be no reparations and no blaming anyone for the war.

The Zionist Jews sent word to the British hold on and they would get America into the war.  Their price was *Palestine to be made into a Homeland for the Jews*.  Wilson makes a campaign pledge to keep America out of the war but this will change after the British get some Americans to board the Lusitania, a British passenger ship, and steer directly it into a demarcated German submarine zone and have the Germans torpedo it.  The Lusitania sinks quickly as munitions put in the cargo hold blew up. Normally it would take a half an hour for a ship to sink but everyone on board could get into life boats.

American newspapers reported that with the sinking, Germans were celebrating the news and German school children were given the day off from school.  Baruch had already talked Wilson into creating the War Industries Board to employ Americans to make weapons to sell to Britain, France, and Russia with the sales financed by the Federal Reserve.

Gold mining engineer, Herbert Hoover, comes to work under Bernard Baruch in supplying food to countries after the war and Baruch grooms him to run as the Republican Party presidential candidate while future President Franklin Delano Roosevelt is vice president on the 1920 Democrat ticket with Al Smith as president and future Prime Minister Churchill is made head of the Colonial Office to set up the division of Turkish territory, make sure Britain controls the oil of the Persian Gulf, set up the

*Homeland for the Jews in Palestine,* America. Wikipedia has posted that America lost 116,700 in the military, 204,000 wounded but its figures for India have only 64,449, which hides one of the most callous and cold blooded action made by any military in the world which the author relates in his book, "The Hypocrites." In effect, America financed the war, provided the weapons, no one paid their debt except Finland. But the American dollar was valued as hard currency with the British Imperial pound.

After Lord Balfour's Declaration that Palestine was to be made into a *Homeland for the Jews*, was hand carried to Lord Rothschild with Balfour's signature which has been called the *Deed to Palestine that was then made into a Mandate for Britain to create, and Winston Churchill was then made Colonial Secretary in charge of creating new nations out of the ottoman Empire.* Churchill was later appointed EX-Chequer of Britain (Secretary of the Treasury in the US) to carry out the Great Depression in the British Empire which lowered values on all property allowing the Rothschild to buy and control the future of the people in the colonies to be little more than slaves of the Rothschild which exists today..

America never joined the League of Nations as it required America's military be put under its control. Congress refused as the Constitution states that only the President of the United States is the Commander in Chief of the Armed Forces. But the League of Nations was used by Churchill to make WW II.

Wilson was against the partitioning of any nation's territory but rendered mute by a stroke. When he was awake he would be photographed in a wheel chair pushed by his young wife but his face does not look like he had a stroke but was sedated and Bernard Baruch ran the country instead of the Vice President.

In WW II as the German Army approached the Suez Canal the British began arming the Jews of Israel to defend Palestine against a possible German invasion. King Saud knew the Jews turned these weapons on Palestinians and refused to accept *Israel's right to exist*. A tall Austrian Jew who had gone to Palestine and learned Arabic, studied the Koran, adopted the name, Mohammed Aziz, went to Saudi Arabia where he loudly proclaimed he was a Jew who had converted to Islam.

On hearing this news, King Saud immediately sent for him and welcomed him as the conversion of Jews to Islam has been dream of Muslims as well as Christians. Saud was charmed and gave him his friendship and Mohammed Aziz wrote a book on the beauties of Islam told Saud, "Look, my friend, if Jews and Muslims lived side by side in Palestine, Jews will see the superiority of Islam become Muslim like I did." Saud was pleased and acquiesced.

Roosevelt met with Saud on an American cruiser in the Red Sea to tell him what the Americans wanted. But Saud did not agree to all or lift Saudi Arabia's ban on Jews entering Arabia. Roosevelt gave Saud an airplane which he used to visit Saudi tribal leaders and Churchill gave him a Rolls Royce which was parked in a garage and never used. Britain and America supported King Saud as head of the Arab world by giving Saudis 75% of the oil royalties and 25% for the Americans.

On August 25, 1941, the Soviet Union invaded Iran from the north and the British the south. Americans were brought in to run the railway and extend a rail line to the Russian border to connect to Russia's railway. The man put in charge of the Iranian Railway was Norman Schwarzkopf Sr. son of German immigrants and a graduate of West Point, veteran of WW I, and head of the New Jersey police. He was credited with solving the kidnapping of the Lindberg baby. He was fired in 1936 in a personality clash with the new governor and worked on radio drama *Gangbusters* before recalled to duty and assigned to Iran to manage the Iranian railway. His son, Norman Schwarzkopf Jr. also attended West Point and led the coalition forces in the Gulf War against Saddam Hussein.

The author wrote how Iran's democracy was destroyed in 1953 by the CIA in 1953 in his book, *The Hypocrites.* That was because Iran's Prime Minister, Dr. Mossadegh, asked for the same deal the Americans gave King Saud head of the Arab World.

which was to make , the British told him, "No, it is 75% for us and 25% for you and it is better than the 16% you were getting."
and has written about the secret atomic war between Russia and China and how the price of oil was raised in 1972 in his story, "*A Star Rises in the East*" in his book *The Age of Disinformation* and why Richard Nixon cut the dollar to float free in the market, the increase in the price of oil, the promises never kept to rebuild Viet Nam, the plan to increase electric power to employ Egypt's millions of workers with nuclear power plants which led to his downfall and his friend George H.W. Bush telling him to resign for the good of the (Republican) party.

A video on you tube reveals that former President George H.W. Bush had issued bonds worth $240 billion to finance *covert* operations to subvert the Russian ruble and bring down the Soviet Union was due to mature on September 12, 2001. $240 billion dwarfs the insurance claims of 9/11. It is far greater the one billion dollars the Mafia/Cosa Nostra paid Ollie North got for his half of the drug warehouse in North Tehran that became known as the Iran-Contra Scandal.

*"I have put away thousands of Americans for tens of thousands of years with less evidence for conspiracy and with less*
208

*evidence that is available against Ollie North and the CIAA people..."* -Former top DEA Agent, Celerino Castillo, from his book, *Powder Burns.*

9/11 is a crime that should have been investigated by the Attorney General, John Ashcroft and Michael Chertoff, they didn't. President, George W. Bush and Vice President, Dick Cheney failed to order an investigation is dereliction of duty.

At 6:45 on September 11, 2001, Israel alerted the CIA that there was to be an attack on the Twin Tower. Information was sent to intelligence and American news media to tell people not to go to the Twin Towers that morning. L. Paul Bremer, CEO of Marsh and McLennon, evidently got the message and changed his staff meeting to a nearby restaurant. But his staff of 12 was waiting for him in his office and they died. Bremer was appointed governor of Iraq May 12, 2003 and served till June 28, 2004. Newt Gingrich said Bremer's performance was called by some a disgrace.

## Supplement to Treachery

Grover Cleveland was the first Democrat elected president since the Civil War. After one term Benjamin Harrison, defeated Cleveland with a smear campaign that Cleveland had fathered a child out of wedlock. Cleveland admitted he had a son outside his marriage and replied that he supported his son and looked after his welfare. Harrison said he was Christian but his grandfather President Benjamin Harrison was a Jew. His presidency was so corrupt that when Cleveland ran against him in the next election and easily beat Harrison. Benjamin Harrison was son of William Harrison who Nathan Rothschild wrote him to put pressure on the Ottoman Sultan to sell Palestine to him. Benjamin became sick on the cold March day of inauguration and died 30 days later, so nothing was ever done. William was the first known Jewish President of America.

President William McKinley was assassinated by Leon Czolgoz who was Polish American who was influenced by Emma Goldman who got $600,000 from a charity *for the victims of the Chicago fire* to finance the publication of her *Die Freie Arbeite Stimme*, the first Communist newspaper in America which she published in Yiddish that is written with Hebrew letters that only Ashkenazi Jews can read. Czolgosz attended a few of her lectures but assassinated McKinley on his own as he said McKinley represented the wealthy who exploited workers. He was executed by the new electric chair a few months after he killed McKinley.

Emma Goldman immigrated to America from Russia with thousands Jews because *Simon Wolf* wrote his *good friend,* President Grant in 1869 to please appoint Daniel Levi Peixotto as US Consul to Rumania to protect the Jews in Russia from Pogroms.

But Peixotto's real purpose was to bring millions of Russian Jews who were also Communists, into America to *replace one million men killed in the Civil War and* 500,000 disabled soldiers in the war who were addicted to morphine given to them in the Civil War. In 1840, there were only 15,000 Jews in America and by 1920 census and over two million Jews and three million by 1930. By 1940 there were 6 ½ million Jews in America, which was the last census of Jews in America as Jews said that a census identified them by area which made it easy for them to be persecuted.

But the percentage of Jews in Congress and top positions in US government and military far exceeds any other ethnic, racial, or religious group. They hold the top positions everywhere from CIA, Military, Social Security, and recipients of major contracts like Kroll of Brazil and Booze, Allen, and Hamilton.

The British used Indian opium in their commerce with Chine as the Britain did not have enough gold for trade with China. Britain declared war on China when the Emperor of China burned the king's opium that resulted in the two Opium Wars with China in which the American navy that assisted the British. Opium was cheaper than gold or silver. In the Viet Nam War, the Vietnamese used opium to weaken the American forces. Zhou en lai called it appropriate for the years of addiction forced on Chinese by the western powers.

In 1915, Samuel Untermeyer visits Woodrow Wilson two months after he is inaugurated president and says, "Mr. Wilson your former neighbor's wife is suing you for forty thousand dollars, *breach of promise*." Untermeyer has the love letters Wilson wrote and she divorced her husband to marry Wilson. Untermeyer offers to pay the forty thousand in return for a favor. The favor is to *appoint Louis Brandeis as the first Jew* to the Supreme Court, and Wilson agrees.

Wilson's campaign manager, Edward Mandel House, is a Jew and introduces him to Bernard Baruch and Henry Morgenthau Sr. They push Wilson to put the Federal Reserve Act before Congress and Senator Nelson Aldrich, of Rhode Island, the son-in-law of John D. Rockefeller, pushes the Federal Reserve Act through Congress.

The Federal Reserve is illegal as the Constitution states only Congress has the right to print the money of the United States and Senator Aldrich committed the crime of nepotism as his father-in-law, John D, Rockefeller was one of the nine Jewish investors in the Federal Reserve Bank and they have the Class A stock with privileges and controls of its operation.

211

Lyndon Baines Johnson becomes president after JFK was murdered, and was elected president in 1964. Johnson, committed treason on June 11, 1967 by ordering planes back from the USS Saratoga that were sent to drive away attackers on the USS Liberty. Johnson decided not to run for re-election.

Nixon came after Johnson and proposed putting nuclear power plants in Egypt because the Aswan dam was not producing enough electricity to expand industry and reduce unemployment and soon after *Deep throat* began talking to *Washington Post* reporters about Nixon covering up the tapes in White House and Nixon was facing impeachment.

Doing things their way created problems for Richard Nixon, Osama bin Laden, and Bill Clinton when he did not initiate a regime on Saddam Hussein but went through the United Nations. President George H.W, Bush did the Gulf War but left Saddam Hussein in power when Russia sent word to leave him alone. George lost the Jewish block vote even though he guaranteed 10 billion in loans for Jewish settlements in Occupied Palestine and lost the election for a second term.

In 1972, President Nixon appointed George H.W. Bush as the first US Ambassador to China. He tells Nixon to resign "for the god of the Republican Party and Gerald Ford appoints George H.W. Bush Director of the CIA when he becomes president, then George loses to Ronald Reagan in the Republican primary and becomes Reagans' Vice President. He was dining with the Hinckley family when John Hinckley is shooting at President Reagan. George rushed to take over the presidency but President Reagan recovered from his wound.

The Lavon Affair, the attack on the USS Liberty in 1967, the 9/11 False Flag. PNAC pro-Israeli (Neo-Cons) advocate a new Pearl Harbor, the sacrifice of 4,000 lives at Pearl Harbor by Franklin D. Roosevelt was to get America into WW II.

In 1962, just after President Kennedy successfully concluded talks with Russia on removing nuclear weapons from Cuba, Prime Minister Golda Meir was meeting with John F. Kennedy in the Kennedy Compound in Miami to discuss the Israel's sabotaging American foreign policy and then praised him for successfully negotiating a peaceful deal with the Russians to remove any nuclear warheads from Cuba and stop keep nuclear weapons put of Cuba. President Kennedy then launched into the need to guarantee peace in the world with international inspection of all nuclear power and nuclear weapons. Golda was nodding her head and when he asked about putting Israel's Dimona nuclear facility under international inspection, Golda Meir, Foreign Minister, said their \would be no objection ben Gurion would not answer.

For the next six months, President Kennedy attempted to get Israeli Prime Minister David ben Gurion permit the Dimona nuclear facility to be inspected. Frustrated with Prime Minister ben Gurion's procrastination he ordered him to sign the non-nuclear proliferation treaty that would allow international inspection. And control of Israel's nuclear bombs. Furious, Prime Minsister David ben Gurion snapped, gave the order, "kill Kennedy," and resigned. John Rich, born in the Philippines to a wealthy Jewish family, James Angleton, head of CIA Foreign Intelligence, Cord Meyer of CIA *Mockingbird* control of US Media, and William Harvey of CIA Assassination Unit, carried out ben Gurion's order and manipulated officials in the Pentagon, secret service, CIA, and FBI who had taken the oath to protect the United States and the life of the president in the oath of office they took and the cover-up continues.

John Lear and Dimitri Khalezov have given evidence that Israel pulled off 9/11. Muammar Gaddafi of Libya made *the same charge* in 2007 so he was annihilated and the new Libyan

213

government has all Libya's oil money and top military weapons and are sending weapons to the ISIS and Taliban.

The UN Assembly specifically ordered the US and UK not to invade Iraq but "wait for the report of the UN Commission that was searching for Weapons of Mass Destruction in Iraq." Mr. Bush and Prime Minister Blair invaded against the UN's order *that made the invasion illegal.*

Former Malaysian Prime Minister, Tun Dr. Mahatir Mohammed, established the Kuala Lumpur War Crimes Tribunal that convicted President George Bush and Vice President Dick Cheney *en absencia* for war crimes of torture, kidnapping, and demeaning treatment of detainees at Abu Graib, Camp Bucca, Guantanamo, and CIA prisons in other countries. The IWCT has yet to respond. But for any decision to be legal, a court of tribunal must have permission from the litigants for a tribunal or court to investigate the charges to see if a trial is warranted.

The International War Crimes Tribunal was set up by the United Nations to investigate war crimes. The United States is a member of the International Court of Justice in The Hague but Congress has yet to ratify it so no American official involved in 9/11, . If George W. Bush, were found to be implicated in covering up and assisting Israel in the killing 3,000 Americans by the *International Court of Justice* the mass murder of 3,000 Americans on September 11, 2001, none of the judges of the court have been given the documents have sent to the court nor has the court office for receiving my documents acknowledged receipt of the documents.

So the author sent the a second set of charges from Germany to The Hague to make sure they would have a second chance to launch an investigation to the charges the author made against Israel for the crime of 9/11. The court has no obligation to respond to a private citizen of the United States. But would it

214

respond if an international lawyer with credentials send a request for an investigation? Most likely not. Israel is above the law.

Israel has gotten Moroccan , Algerian, and Tunisian Jews the status of *Holocaust Survivor* even though none were imprisoned or mistreatment by Germany. The fact is that all were still governed by Vichy France as Germany left them colonies of Vichy France.

Benjamin Freidman was a top Jew who liaison for Henry Morgenthau Sr. who wanted Wilson to appoint him a cabinet position. Instead, President Wilson made him US Ambassador to Turkey *to protect the Jews of Turkey.* Morgenthau used his position to write reports about Turkish genocide in Armenia which he sent to the *New York Times* to inflame hatred against Turks and serve to gain support of taking Palestine from Turkey to make a *Homeland for the Jews.*

In 1964, Mr. Friedman made this speech from the Lincoln Hotel in Washington. He personally knew every president from Woodrow Wilson to John F. Kennedy and said:

"Do you know what Jews do on the *Day of Atonement* that is so sacred to them? I was one of them. This is not hearsay. I'm not here to be a rabble-rouser. I give you facts. On the Day of Atonement when you walk into a synagogue, you stand for the very first prayer that you recite. It is the only prayer for which you stand. You repeat three times a short prayer called the Kol Nidre. In that prayer, you enter into an agreement with God Almighty that any oath, vow, or pledge you may make during the next twelve months shall be *null and void*. The oath shall not be an oath; the vow shall not be a vow; the pledge shall not be a pledge. They shall have no force or effect. How much can you depend their loyalty? You can depend upon their loyalty like the Germans did in 1914. We are going to suffer the same fate as Germany suffered, and for the same reason." - Benjamin

215

Friedman, was a loyal and true American patriot unlike those involved in getting American into WW I, WW II, and 9/11. WW I was to destroy the Ottoman Empire, get Palestine the Sultan refused to sell, and WW II was to save Britain that held the Mandate of making a homeland for the Jews in Palestine.

Left out of NIST official report was the military exercise that were to test the defense procedures for an air attack on the United States. The test was being conducted at the same time of 9/11 and was under control of Vice President Dick Cheney.

April Gallop, an Administrative Specialist for SCI in the Pentagon had been on maternity leave. This was her first day back to work and brought her two month old son with her. She and placed him under her desk and would later take him to the child care center. She said as she touched the first key on her computer a bomb went off. She picked up her baby son and headed for the hole that a Boeing 767 was supposed to have made. She walked past computers that had small fires but there was no luggage or debris from an airplane. When she got outside, two men rushed to support her as she was ready to collapse. One of the men turned and went through the hole she came out of and never returned. April was visited in the hospital a few days later by a man dressed in an army officer's uniform who asked, "What happened?"

April Gallop told him about the explosion that blew a hole in the wall of the Pentagon. The man wearing an army officer's uniform who never gave his name said, "No. You saw a plane that made that hole. There was no bomb."

She searched for months to find a lawyer to sue Vice President Dick Cheney, Donald Rumsfeld, and General Richard Meyer. All the lawyers she approached refused to take her suit until William Veale accepted to represent her. His suit was rejected and a fine of $15,000 levied against Mr. Veale for filing a *frivolous* suit. One of the judges was John *Walker, a cousin of President George W. Bush.*

Phillip Marshal was a CIA pilot for many years and also flew commercial passenger planes. He wrote several books before he became interested in the 9/11 Truth Movement and he has divulged that Evergreen International was used by the CIA, and flew charter flights for the CIA. Evergreen was based at Pinal Air Park, Marana, Arizona and could paint a Boeing 767 like commercial airliner for a Black Ops operation and had done previously for the CIA. Phillip knew his knowledge might result in his death. He was shot in the left temple but he was right handed. The Sherriff's called it suicide and a double murder of his teenage daughter and son. Classmates from their school went looking for them and said the door to the house was wide open. Sheriff's office ordered the crime scene cleaned destroying all evidence of a murder while the FBI removed computer, file cabinets and papers.

Were the United and American Airline planes leased from Evergreen? Evergreen went bankrupt with its last flight December 3. 2013.

Barbara Honegger learned that the plane from Dulles scheduled to fly to Indianapolis was tracked by radar to southeast Ohio. The other two planes that *hit the twin towers* were tracked by radar to Cleveland, Ohio. Two years later an air controller at Cleveland identified one the planes as landing at Cleveland by its identification numbers.

CSI is an Israeli security firm that does the security for Logan and Dulles International as well as the security of Schipol International at Amsterdam where the shoe bomber and jockey underwear bomber departed for flights to the US.

Debra Palfrey supplied beautiful *escort girls* for powerful men in Washington. On the Steven Jones Talk Show she said

217

three times that she would never commit suicide. (She was planning to give names of powerful men who used her escort girls at $300 for 90 minutes. She was found hung by a nylon cord in a shed on her mother's property in Florida

Al Carone worked for CIA Black Ops team and said Mike Harrari could get into any US military base anywhere and anytime without prior notice. Al was involved in the massacre of a village of Northern Mexico and became despondent over this wanton killing babies, women, and children. He wanted to quit the CIA, but died in an Army Hospital of radiation poison like Yasar Arafat.

Michael Ruppert was head of LAPD's narcotics division and resigned over LAPD's blocking his attempt to prosecute drug trafficking by CIA, the Contras, and Ollie North, who was exposed by Ayatollah Montesseri for billions in drugs for weapons Iran needed to fight 'Saddam Hussein of Iraq. Ayatollah Khomeini put him under house to silence him as his son and two other young men were involved and they accepted six million Canadian dollrs in bribes. Ruppert documented his work and wrote, *Crossing the Rubicon*. Hounded and bankrupted by the IRS, FBI, and CIA he is documented as a true suicide while Deborah Palfrey and Phlllip Marhall and his son and daughter were murdered..

A top characteristics of the sociopath is the lack of sympathy for others. Any show of sympathy is feigned much like Larry Silverstein's telephone call to the NYC Fire Marshal where he says, "You know there's been such a terrible loss of life, maybe the smartest thing to do is pull it."

On September 11, 2001, explosions inside the twin towers demolished stairways and exit doors were chained to trap people inside and marked for death. Compare that with the joy of those two men dancing on the stage *in the Guangzhou Hotel, giving themselves hi-fives,* or *the five dancing Israelis giving themselves hi-fives and having their pictures taken, with*

218

*the smoking towers in the background,* and the *dancing Mastermind, Colonel Mike Harrari. Sociopaths are criminally insane and this condition is incurable.* Sociopaths should be isolated from human society and interred in a mental facility for their lives as they deceive to manipulate others to commit murder and other crimes. They have no feelings of guilt or remorse for bringing harm they have caused others. Do they kill? Of course and steal. They pose as experts and can tell the most convincing lies. The live on the edge of the law as this excites them the most and otherwise they feel bored. They want to be noticed, to be a celebrity. Their condition is incurable.

9/11 was not just a crime committed only by Jews or Israelis. There were gentiles, agnostics, Israelis-Americans, news media, celebrities, professionals, police, and people holding a government position. But the torture of detainees that these *in power* knew to be innocent are crimes where their identities are known and they should be prosecuted. The American Psychological Association has not decided what to do with its members who trained, gave legal advice, and participated in the torture of detainees. The CIA had its own facility, Delta, where it requested certain detainees for more torture and the basic hardliner ISIS leadership was formed. But they violate the very principal of Islam that states all humans are brothers and sisters. Of course there are thieves, perjurers, liars, and killers among all nations. Those are separate issues dealt with by criminal law. But to kill others because of religious, ethnic, national identity, or race is a heinous crime and the leaders know this but persist because their minds are distorted. And those who have done this warped mind through torture must be prosecuted and isolated from society for life.

9/11 was an act of war against the United States by Israel as well as the murder of President Kennedy and the attack on the USS Liberty to 9/11, all evidence has been sealed.  The murder of 3,000 has shown how much Israel values human life and  the friendship of Americans who have given billions annually to Israel and guaranteed the protection of Israel, and blocked charges of crimes against humanity committed by Israel with America's veto power in the UN Security Council.

The story of how the Granit missiles landed up in America involves a former Russian spy, Viktor Bout who is Dimitri Khalezov's friend.  Bout began trafficking in weapons the same way and about the same time as Ollie North began trafficking in illegal weapons from the Reagan White House.  Ollie was protected by the CIA and DEA.  The CIA and Mossad worked together in the Golden Triangle which became big in the Vietnam War with opium and heroin put in the body bags of American soldiers killed in Vietnam that were sent to America for burial and that got the drugs past Customs.

On December 25, 1991, Russia was broke and needed exports and Victor Bout became important to Russia to obtain foreign currency.  Viktor did accept cocaine as payment for weapons as did Israel and the US.  Israel ships its weapons by its fleet of merchant ships registered under the Liberian Flag of Convenience and Republic of Panama.

A Mossad the author knew personally in Honduras, was undercover security on Israeli ships that transported weapons to Asia, Africa, and Latin America, FARC and FLMN.  Anyone on the ship who talked about the cargo and its destination would find himself in the Ocean far from land.

When the Government of Sri Lanka wanted to buy weapons from the United States to fight Tamil rebels in the north, they were told to buy from Israel.  Israel supplied and trained both Sri

Lanka and the breakaway Tamil Tigers. It was a long war and the Tamils lost so many men that some Tamil Tiger battalions were more than fifty percent women.

Israel's biggest industry is manufacturing weapons but its export of illegal drugs may be bigger than its export of weapons. Israel's sales of illegal drugs undermine the Euro and dollar and its nuclear weapons are a drain on its economy as it costs around two hundred million dollars to make one atomic bomb. Israel is the first nation to make mini-nukes. .

Dimitri Khalezov revealed that mini-nukes were used to bomb the US Embassies in Tanzania and Kenya, car bombings in Iraq and Afghanistan, the Bali bombings, Nogal Towers in Saudi Arabia, Oklahoma City Bombing, the 1993 bombing of WTC North Tower, and 9/11 Twin Towers, all to be blamed on fanatic Muslims. Dimitri says 50 kilos of U235 were planted on the FARC camp in the Ecuadoran jungle to blame them for the mini-nuke bombings in Bogota. The mini-nuke in the Granit missile warhead that hit the Pentagon did not detonate or April Gallop would not have lived. Mini-nukes are plutonium is highly fissionable U239 and mini-nukes were first invented by Israel. It takes heavy hi tech industrial machines to make U239 that goes into mini-nukes which can never be a cottage industry in some Arab village.

Dimitri says Netanyahu's claim that the UN inspection cannot keep Iran from making an atom bomb is to fool gullible politicians and mid-level security people. UN International Inspection can and will work. In the author's opinion, all nations with nuclear technology: Israel, the US, China, Russia, Pakistan, etc. must submit to it or dismissed from the United Nations and sanctions imposed.

Marsh and McLennon occupied eight floors in the North Tower and a girl is seen waving from the hole in those floors occupied by Marsh and McLennon and the hole was supposedly made by a plane.   Melissa Doi was on the 83 floor and called 9/11 on her mobile.  She pleaded for help as she and five others were on the floor as the smoke was too much they had to lay on the floor to breathe.  She could not take pain and was almost screaming.  That heat was coming from explosives ignited to plow out a hole to look like a plane hit the building.  But a girl waving from that hole was videotaped waving just22 seconds before the building copplased.

The author has introduced drug trafficking by the CIA and Mossad into this book because of the US extradition and trial and incarceration of Viktor Bout for weapons trafficking.   His indictment by John D. Negroponte who was US Ambassador to Honduras at the time Ollie North, the Contras, and Juan Ramon Matta Balesteros were involved in drug trafficking.  The reader can read the honors heaped upon Negroponte and wonder how these people in these organizations can speak, write, and deliver such accolades should have the ordinary citizen wonder at the education, his appointment to high positions were merited?

The very idea that Muslims could or would export terrorism to America started with the 1993 bombing of the WTC North Tower with mini-nukes and the FBI sunder Louis Freeh framing blind Sheik Omar Rahman who worked for the CIA in recruiting Mujahedeen to fight the Russians in Afghanistan.   The CIA Mujahedeen went to Bosnia to defend the Muslims of Bosnia. Al Qaeda does not exist.

Israel pulled off 9/11 and Prime Minister Ariel Sharon and Mike Harrari were the war criminals who were behind it.  The reader will see that the author includes Franklin Delano Roosevelt, Woodrow Wilson, Winston Churchill committed war criminals.  Joe Kennedy Sr., the US Ambassador to Great Britain, told Roosevelt to stay out of the war in Europe.  Was his son, Joe Kennedy Jr., deliberately killed in Operation Aphrodite?  The

222

story that Joes Kennedy Jr. was killed at the end of WW II is false as shown by the records of Operation Aphrodite, unsealed in 2008, show he was killed in 1943.

The Bush White House Legal Advisor, Alberto R. Gonzales defended the use of torture by saying the Geneva Convention was *outdated and unrelated to current events.* He was made Attorney General of the United States and now teaches international law at Harvard. He is an example of the corruption of the United Nations which replaced the League of Nations and has legitimized all the actions of Israel which have broken all of the Ten Commandments of God that are basic to Muslims, Christians, and Jews. Arthur Kessler estimated 82% of the top Communists of Russia were Jews.

The IWCT and International Courts of Justice are not doing their job to isolate criminally insane sociopaths from the rest of humanity and must be cleaned out or replaced. Another factor is that nations must bear more responsibility to scrutinize those who are selected or elected to serve.

# Craters in the Bedrock

The craters in the bedrock of the twin towers were made by mini-nukes placed in the basements of the towers that were detonated on September 11, 2001. The crater shown on page 48 is 100 meters (328 feet) wide and 27 meters (88 feet) deep and carved out of the bedrock on which the twin tower stood. The debris from the towers filled the crater otherwise the debris would have been seven stories high. The craters are not the only proof that nuclear weapons were used to bring the buildings down. There were electro-magnetic impulses recorded that are a sign of nuclear detonations. This photo on the cover of

"Treachery" is proof 9/11 was not done by Arabs intent on destruction with hate for America but with hi-tech mini-nukes made by Israel.

The upper mantle of liquid Iron of the Earth's core thermal movements, spur the Earth's rotation and its magnetic field holds the Earth's atmosphere to the Earth, (April 26 issue of Science), .

At the plasma temperature of the surface of the sun, electrons fly away from the nucleus to produce plasma which is one of the states of the four states of the elements called plasma. The earth's crust is what protects us from the high temperatures of the liquid mantle. The mini-nukes in the basements of the twin towers vaporized the bedrock on which the twin towers were built.

September 11, 2001 was the day America's greatest friend in the Middle East nuked America and killed more than three thousand 3,000 Americans (how many were vaporized in the in the 338 elevators), and Rabbi Dov Zakheim stole 2.3 trillion dollars.

The steel support girders of the twin tower were placed on bedrock which vaporized at 7,200 F. Bedrock is about 50% oxygen which will vaporize while steel will only melt. As the author gave the information that the Czar bomb was 1,570 times more energy than the energy of Hiroshima bomb that left Japanese with skin hanging like cloth from limbs like man William Rodrigues rescued testified from an elevator shaft shows 9/11 was nuclear.

The Earth's upper mantle of its core is 1,000 degrees Celsius hotter than previously measured. The precise measurement was to measure the exact temperature and exact moment iron melts. The French scientists doing the research used a new X-ray technique that takes measurements faster than before. Iron samples compressed in the laboratory lasted only a few seconds thus making it difficult to determine if the iron is still a solid, or starting to melt. The new technique used the diffraction occurring when X-rays or other light hit an obstacle and bend around it. By sending burst of X-rays at a sample of iron, scientists were able to pinpoint that temperature and put the melting point of iron at 4,800 C at a pressure 3.3 million times atmosphere pressure at sea level. Scientist extrapolating their measurement of the boundary between Earth's inner and outer core is 10,832 F. +or - 930 degrees Fahrenheit under 3.3 million atmospheric pressures..

The research was conducted by CEA French national technological organization and French National Center for Scientific Research and European Synchronization Radiation Facility (ESRF)

In 2013, Israel's news media revealed that Israel planned a nuclear attack on Iran's atomic power facility but it was not carried out as Israel's two top generals were not sure how far along Iran's nuclear energy program had developed and nuking Iran might have serious repercussions. The craters in the bedrock below the twin towers showed that the mini-nukes used to melt the steel pylons that gave the towers the strength to withstand hurricanes and multiple airplane hits vaporized the bedrock. Today's H bombs are fifty times more powerful than the one used on Hiroshima and Nagasaki. Two nuclear bombs dropped on New York City built on an extinct volcano has the Palisades across the Hudson River which is a sill.. That is fault filled with extruded molten volcanic rock shows the complex geology of New York City. Two or three mishear bombs on it would affect the East Coast and New England.

The Earth's crust is made up of 12 tectonic plates separate by fault lines and float on molten rock of the interface. .A fusion (H) bomb hitting a fault line just might release the pressures on that molten rock would be many times greater than Global warming.

All nations with atomic weapons and nuclear energy must put these programs under international inspection and control. Work for arms limitations and reduce the number armed forces and put all intelligence agencies under international control.

On September 11, 2001, has 2,780 known dead but how many were vaporized in the elevators? Hundreds of human beings night have been vaporized or cremated alive in the 348 elevators. As the twin towers collapsed a plume of white vaporized rock rose up in the center of the building, reached about 1,000 feet where north winds blew it out to sea. Those bodies that were vaporized would in the elevators would not be a part of the. 6,000 pieces of human flesh recovered and identified or of the 10,000 unidentified pieces.

Mike Harrari, Zionist politicians, the CIA, FBI, Interpol, Secret Service, Pentagon, all armed forces of the United States, the Federal Reserve, NATO, UN Security Council, and the News Media of the western nations were all involved in 9/11.

The author has been in six earthquakes in his life: Iran, California, Honduras, Costa Rica, and Colombia. The Earth's crust is more fragile than the average person thinks.

The greatest earthquake in recorded history was the New Madrid earthquake of 'SE Missouri, 1812. The Mississippi River rolled backward, the land rose and fell like waves in the ocean, and when the Mississippi River came back it cut a new river bed and the old river bed formed the Reel lakes of Southern Illinois.

The Earthquake that shook Anchorage. Alaska came from a depth of 147 kilometers below the Earth's surface. The March 27, 1964 Anchorage earthquake was 9.6 magnitude and ruptured 600 miles of fault line, raised the land 60 feet, and produced a tsunami.

Managua, Nicaragua was built next to an ancient volcano that is filled with beautiful blue water that comes from cobalt which is poisonous to humans and plants. Unknown to Nicaraguans, Managua had many fault lines under it. The night of the big earthquake, the air was still. The sky was red and loud noises were coming from it not the Earth and the ground began to shake. The Central Bank building dropped straight down and other buildings cracked, and fell. A whistler effect in the upper atmosphere tells seismologist that an earthquake will happen soon but not where.

feet thick 12,000 years ago. The volcano on the Island of Manhattan, New York City, was scraped off and Long Island is debris left when that prehistoric giant icecap melted and retreated.

Iceland, North Island of New Zealand, and Yellowstone Volcan0 Park are the only places in the world with geysers. Water seeps down into the fault line, and at the boiling temperature of water expands 999 times which cannot be compressed.

Yellowstone's prehistoric massive flow of lava created a dam behind which the world's largest lake was formed covering more than three Canadian Prairie Provinces. When the dam broke, the water from the lake rushed through and created the bad lands of western Oregon, Washington states, and the Colombia River gorge.

The Hawaiian island were created by a hole in the Earth's crust on the bottom of the Pacific Ocean. The big Island of Hilo sits on this hole the rest of the islands moved westward and it has continuous lava flows adding more land and destroying villages.

Ayatollah Khomeini refused to renew the agreement for foreign oil companies to exploit Iran's oil, so the CIA tempted Saddam to attack Iran and annex Iran's oil rich Khuzestan Province that would make Iraq the biggest oil producer and head of the Arab World. To prove they were his friend, America removed all the hi-tech weapons Richard Nixon allowed the Shah to buy that made Iran the 3rd most powerful nation of Asia after Russia and China.

When Ayatollah Khomeini took power he invited the PLO and Yasar Arafat to Iran, then an invasion of the American Embassy. But Foreign Minister Ibrahim Yazdi went to the embassy with an Imam and talked them into going home. Then Prime Minister Mehdi Bazargan and Dr. Yazdi wanted Mr. Carter to release the 25% of Iran's oil royalties frozen in American banks that were part of the agreement CIA Richard Helms had the Shah sign after the CIA overthrew the democratic government of Iran's Prime Minister Dr. Mossadegh. Mr. Carter put his foreign advisor, Zbigniew Brzezinski in charge of dealing with Iran and he offered to meet with them but said Iran was too far and to meet him in Algeria. As they flew to Algeria and the Mossad carried out the second invasion of the American Embassy. And on their return to Iran, Khomeini fired both Dr. Bazargan and Dr. Ya8zdi and Abdolhassan Banisadr was made the first President of Iran and Ghoptzadeh was the Foreign Minister.

The CIA gave the Mossad a list of American and Russian agents in Iran, and the names of Iranian pilots waiting to overthrow Khomeini. Number on the list was Banisadr and number Two was Foreign Minister, Ghoptzadeh. Banisadr escaped to Paris with Masoud Rajavi, brother whose brother headed the Mujahedeen Khalk (80,000 Iranian Communists) that the CIA financed, trained supplied and then turned over to Saddam Hussein to fight Iran.

Ghoptzadeh didn't get the message and was executed as were hundreds of Iranian pilots, Russian and American agents. Ayatollah Khomeini had invited Yasar Arafat and the PLO to come but then kicked them out and invited the Mossad back.

## Total Cost of 9/11 from 2001 to 2009

| | |
|---|---|
| Estimated uninsured losses from 9/11 | 35,600,000 |
| Cost of building the twin towers in 1972 | 37,000.000 |
| Cost of building the whole WTC: | 90,000,000 |
| PANYNJ refunded Larry Silverstein | 98.000.000 |
| Israel reimbursed by Reagan for Lavi fighter | 500.000.000 |
| Initial installment of Larry's 7 billion | 616,000,000 |
| Value of gold stolen from vaults | 850,000,000 |
| Insurance payment to PANYNJ for loses : | 950,000.000 |
| Industrial Risk Insurers on building 7 | 861,000,000 |
| Silverstein's insurance claims for rents | 7,000,000,000 |
| Bonds issued by George H.W. Bush | 240.000,000,000 |
| Estimated salaries of Homeland Security | 300,000.000,000 |
| Nov 28, 2011, *Forbes,* 2nd Bailout | 16.000.000.000.000 |
| US Federal revenues for 2001 | 1,990.000.000.000 |
| US Government expenditure for 2001 | 1,860,000.000.000 |
| Money Dov Zakheim stole | 2,300,000.000,000 |
| Total losses to US Economy | $ 20,401,037,600,000 |

***The message in this book is that all nations having nuclear weapons and nuclear power must have international inspection.***

*Any nation not responding to this requirement must be removed from the United Nations and isolated.*

Israel's biggest industry is making and selling weapons to both sides of a conflict and it is the biggest arms smuggler in the world. Israel's economy depends on sales of drugs, aid money, and investments from overseas. Israel wants to keep America in constant war to keep millions employed in producing weapons as the US is Israel's its biggest customer. Israel spies on the US military industrial complex to obtain the latest in American weapon technology.

Johnathon Pollard was born in America but spied on America for Israel and was sentenced to life in prison. He was released November 20, 2015 after serving 30 years. Sheik Omar Rahman was brought to the US by the CIA to recruit Muslims to fight the Soviet Union in Afghanistan. He was framed by the FBI and CIA in the 1993 bombing of WTC North Tower to turn Americans against Bosnian Muslims as Israel feared it would have Americans turn pro-Palestinians. Rahman and followers were innocent.

One of the demands of the US supported ISIS terrorists in northern Syria and Iraq is the release of Sheik Omar Rahman. So why do the nations of Israel, the US, Turkey, and Saudi Arabia demanding Bashar Assad's removal from power? It is to legitimize Israel's illegal annexation of Golan Heights and create a pipeline for Gaza's gas field in the Mediterranean to connect with the Russian gas pipeline to Europe. So why has Vladimir Putin whose Gazprom holds the rights to exploit Israel's and Gaza's gas fields come to the aid of Bashar Assad?

Mordecai Vanunu, an Israeli Peace Activist, was born in Morocco. His family moved to Israel in 1963 after hostility to Jews began in Marrakesh, Morocco where he was educated. Conscripted for military service in Israel, Gohe applied for pilot training but trained as a sapper. After discharge from the IDF, he traveled Europe and worked in many small jobs before returning to Israel to work at the Dimona nuclear weapons plant. He was schooled in physics, math, chemistry, first aid, fire safety. He planned to travel but before leaving, he took 51 photos of the interior of the Dimona nuclear plant. After traveling to SE Asia and then to Australia where he decided to immigrate. Mordecai converted to Christianity and joined the Australian Anglican Church and thereby became a *Goyim*.

In Australia he met a Colombian and told him about the Israeli reactor. The Colombian said his story was worth millions and contacted the *Sunday Times* with his permission. But *Sunday Times* people were wary of being duped, so he flew to Lohdon where the *Sunday Times* had Theodore Taylor and Frank Barnaby listen to Mordecai's detailed description of *lithium-6 preparation* which is essential the production of *tritium* used in *fusion-boosted fission bombs*. Mordecai gave a production rate of 30 kg a year with 4 kg per weapon. Barnaby and Taylor concluded Israel might have 150 nuclear warheads. Both experts agreed Mordecai's account was factual and correct.

The Mossad learned about Mordecai from Robert Maxwell and began to focus on him. They learned Mordecai longed for female companionship and assigned Mossad agent, Cheryl Bentos, to seduce him as an American named *Cindy*. This kind of operation is called a *honey trap* which is practiced by the CIA and may have been used to trap Julian Assange in Sweden to humiliate or silence him and get control of Bradly (Chelsey) Manning's files he sent to Assange. Anyone with information that will show Americans are involve in War Crimes will be assassinated or put in prison to be silenced.

Mossad Cheryl persuaded Mordecai to fly to Rome for a holiday with her. After landing in Rome, she suggests they take a taxi to an apartment in the old quarter of Rome where upon entering the apartment, three men jump on him, inject him with a paralyzing drug, and a white van hired by the Israeli embassy arrives that night and he is carried to it on a stretcher. The van drives out of Rome and down the coast to a speedboat which takes him to the Israeli electronic spy ship, *Noga,* anchored in international waters and then sails directly to Israel once Mordecai is on board.

October 6, the *Nogo* anchored off tsrael's coast between Tel Aviv and Haifa and another boat took him ashore where he was taken into custody and interrogated by the Mossad

Mordecai was tried in secret, convicted of espionage and sentenced to 18 years in prison. In 2004 he is released with restrictions but he reveals President John F. Kennedy was murdered by order of Israel which confirms the account by ex-CIA pilot, John Lear, who said John Rich and James Angleton were given the assignment. Those involved with Rich and Angleton were: Cord Meyers (in charge of CIA's *Mockingbird* that controls the News Media). Mary Pinchot Meyers divorced Cord Meyers after their nine year old son was killed by a hit and run car in front of their house. Mary's friends say she went a little wild and had a love affair with President Kennedy and was then murdered a year after JFK was murdered while jogging on the Potomac-Ohio canal towpath.

James Jesus Angleton broke into Mary Pinchon Meyer's house after her death and was looking for her diary which detailed her love affair with President Kennedy. E. Howard Hunt told his sons about his involvement in the murder of JFK and gave the names of Frank Sturgis and William Harvey (who was in charge of the CIA's assassination unit). His widow was interviewed in a video on you tube as the widow of a Secret Service agent and refers to Jack and Jackie as *scum.* What do all Americans think of men paid big salaries to protect the President but betray him and take orders from Israel to kill the president on November 22, 1963, manipulated by John Rich, James Jesus Angleton, and William Harvey?

Four years after ben Gurion gave the order to kill Kennedy, Israel attacked and killed 32 American sailors and wounded 174 others on the USS Liberty. Again people in the Pentagon and Naval Intelligence were involved. Former naval officer, President Johnson, knew President Kennedy was going to be killed and covered up Israel's attack on the Liberty.

Mordecai was on the list of Nobel Peace Prize candidates since 1987. But in March 2009, Mordecai wrote to the Nobel Peace Commission he wanted his name removed from the list of candidates as Israel's President Shimon Peres of Israel, who was in charge of Israel's nuclear weapons program, was on the list of Nobel Peace Prize laureates. Mordecai deserves to be made an American citizen and returned to America. Pollard who spied on America was given Israeli citizenship and given a heroes welcome..

Andrew and Leslie Cockburn lived in Israel and said the SCUD missiles fired by Saddam Hussain into Israel during the Gulf War did create fear and damage.. The Patriot missiles given to Israel by the US to protect Israel proved to be more dangerous to Israel.end missiles with small nuclear warheads that Israel fired into Iraq left radiation in the air and it has been documented that hundreds of babies born in Bagdad born after the Gulf War showed deformities that were proof civilians were targeted. A US intelligence officer said that missiles from Israel landed near or in Baghdad with the mushroom cloud formation attesting Israel committed a war crime.

In 1967, Israel dammed the Jordan River, seized the springs below Golan Heights, tapped the springs that provided Gaza with water and turned all that water into Israel. Israel then bombed the two dams the King Hossein built on two tributaries of the Jordan River to supply Amman and Damascus with water. That act of destruction does not appear in the News Media.

Andrew Cockburn revealed that in 1967 Israel was poised to attack all its neighbors but held off until its Mossad came back and said the Americans agreed to it. Israel's Foreign Minister, Abba Eban, then flew to Washington and told President Johnson that Israel was afraid of the angry Arab rhetoric and was going to make a preemptive strike on all its neighbors and Johnson gave Israel asphalt shredding bombs to prevent Egypt's air force from taking off to defend Egypt.

**Andrew and his wife** Leslie who is also a journalist have lived in Israel. In an interview in 1996, he revealed that Mike Harrari, head of the Mossaf in Centrl America supported and given weapons the Contras in Nicaragua.

This was the result of a **Mossad** agent in Honduras writting report that the Sandinistas were building airfields in Nicaragua to support an air invasion by Fidel Castro and take over Central America. But the airfields were built by ranchers and sawmill operators to fly to their operations as Nicaragua were bad.

The real reason for Israel supporting the Contras was that Jews of Nicaragua were angry with the Sandinistas were nationalizing land given to them free by Anastacio's Somoza's father in the 1930's and they needed to get hard currency by exporting the bananas from the banana plantations the Jews made of that land.. ..

On Jue8, 1967, 15,000 Egyptians soldiers in Gaza put up little resistance to the Israeli Army and surrendered. The Israeli commander in Gaza ordered 1,500 Egyptian soldiers *executed.* Some had their hands nailed to palm trees and were used as human targets. Israel invaded Syria two months before and took

Golan Heights and then planted nuclear mines to deter Syria from ever attempting to retake it.

In 2014, Oliver Stone said to Bill Maher on Maher's TV show, "Israel's interests are not our interests." Maher disagreed and said, "Israel's interests are American interest," and went on to say American news media is not pro-Israel. But he also said he is a strong supporter of Israel and that the US has an *obligation* to defend Israel and claimed that10,000 rockets were fired from Gaza into Israel. Where did these come from as Gaza is completely surrounded an land and with a naval blockade?

Leuren Moret, an investigative journalist, said Israeli agents crossed over I onnto Gaza and fired rockets from Gaza that killed no one or destroyed any property. Israel then sent tanks and fired artillery shells that burst in the air like those artillery shells Russia's Yeltsin gave Serbs in Bosnia used to fire into the largest market of Sarajevo cutting people to pieces.

Israel has annexed 70 % of Gaza and wants to annex the rest as well as annex all Israeli settlements in the West Bank that comprise over 80% of the West Bandk. Israel wants to exploit Gaza's off-shore gas field that that is worth billions. US State Department was supporting Israel and helping to clear the way for Israel to exploit its gas field and the gas field off Lebanon.

Dr. Michael Scheuer was the desk officer on Osama bin Laden in the CIA at Langley and stated that Osama Bin Laden was classified as a *resistance fighter.* They fight against armies and do not do terrorism. Scheuer says Osama bin Laden was not part of the Al Qeda Movement but an inspirational figure. Bin Laden was welcomes in Gulf States because the CIA had made him a hero for helping the Mujahedeen fight the Russians in Afghanistan.

The Mujahedeen were taking heavy losses until Osama bin Laden came to Afghanistan withb his heavy earth moving machines as the request of the CIA and made warehouses, living quarters, hospitals, and fortresses in the mountains of Afghanistan. The training areas were built by the CIA so the US knew where to bomb them and claim they were made by bin Laden..

Dr. Scheuer explains *terrorist Muslims* are those who seek revenge. The collateral deaths by US mercenaries and soldiers was in the information Chelsea Manning gave Julian Assange disclosed torture, kidnappings, and detentions centers where anyone may be held for years without any legal charges made against them which is work the War Crimes Tribunal is to do.

Scheuer says America must stop intervening in the world and all these wars America has been involved in are not worth one American life or one dollar.

Turkish investigative journalist, Ergun Poyraz published his book, *Children of Moses,* and accused Recep Tayyip Erdogan and his wife of being secret Jews who collaborated with Israel. Poyraz was arrested in 2007, and six years later sentenced to 29 yers on August 5, 2013, for sedition.

As Jesus and Abraham, were both are called Jews and were called Prophets in Islam, being a Jew is not crime. But like the man who came Medina which is where the Prophet Mohammed is buried in the huge Mosque, this mn bought a house near the Mosque as he said he wanted to be near The Prophet. But in reality he was digging a tunnel to the Prophet's grave to steal his body. He was caught and executed many years ago when Saudi Arabia of today was under Turkish rule. The Prophet's body was exhumed and a strong iron chain were put under and around the sides, and covered the top to prevent any future attempt to defile the Prophet's tomb. The Prophet has never been deified and Muslims say he is only a man but a man who talked with God. All decisions in Islam are to be made by everyone voting. When there came a time of setting camp before the sun set, the group called for one spot and the Prophet chose another. When asked if Allah had spoken to him about this, the Prophet said no. So the vote was taken and the majority ruled differently. The Prophet Mohammed is revered but not Worshipped and there is no clergy set up in Islam as there is in the Christian sects. Such thing as having Imams call for battles and the executions ordered ISIS Imams is completely against Islam.

The Jews of Medina manipulated the pagan Arabs into attacking the Prophet and his followers. However a convert from Persia (Iarn) by the name of Salman told the followers to dig a deep ditch and that deep ditch protected them from a charge by the pagan horsemen who were ready to kill the Prophet and his followers.

Another plot to kill the Prophet by the Pagan Arabs had his cousin, Ali, cover himself with the Prophet's green robe and lie on his bed while the Prophet escaped. When the assassins discovered that it was Ali laying under the robe, they sent for trackers to follow the Prophet's trail through the night. The Prophet had taken cover in a hole where the trackers had discovered. But as a spider spun a web covering the entrance they turned and followed other tracks and this is considered a miracle attesting to the protection Allah.

Turkey's military executed President Adnan Menderes in 1960 for changing the Muslim call to prayer from Turkish to the original Arabic. It was based on the Peace signed by Mustafa Kemal Ataturk in Switzerland that allowed the creation of a Turkish Republic and creation of Christian and Jewish schools but no Muslim schools were to be allowed. So when one man claiming to be an expert on Turkey claimed that Muslim schools were established all over Turkey and were teaching extremism, the author challenged his allegations and alleged expert disappeared from the internet..

Dr. Mohammed Morsi, head of the *Justice and Freedom Party* and was the first President of Egypt after the *Arab Spring* ended Hosni Mubarak reign. When he said he would work for the release of the blind cleric, Sheik Omar Rahman, sentenced to life for the 1993 bombing of North Tower, the Egyptian military ousted him.

Terrorists that gunned down British tourists on a Tunisian beach, did not receive news media attention as much as the November 2015 attack on a Paris humor magazine after which French Muslims were attacked. The TV program *Sixty Minutes* interviewed the French police who said the killers were common criminals looking to get some fame and were not extremists but celebrity seekers.

Non-PKK Kurds found food and weapons from Israel in territory they have freed from ISIS control. In late October 2015, the Iraqi Army captured Israeli Colonel Yuel Oulan Shabak with Takfiri IS rebels. Israel treats both ISIS and ISIL wounded in Israel. But why is Syria's embassy in the United State closed?.

Israel has bombed Syrian Army units in the suburbs of Damascus and Bibi Netanyahu has asked Mr. Obama to recognize Israel's illegal annexation of Golan Heights, seized two months before Israel attacked on all its neighbors June 8 1967. Netanyahu says Golan Heights is part of ancient Israel which is a lie as the springs at the bottom of Golan Heights have supplied Damascus with water for over five thousand years, which is before Israel ever existed. The US Congress that gave him a standing ovation yet November 7, 2012, Netanyahu said, "We will squeeze all we can out of the United States and it can dry up and blow away."

The question is not merely "Is Israel a menace to America" but is Israel a Menace to the world?

Erdogan's Turkish intelligence killed Lebanese-American journalist Serena Shims who discovered that ISIS volunteers were taken across turkey's border in NGO trucks. On October 19, 2014 near the Syrian border city of Kobani, a huge cement truck crossed the road to hit her car head on. Serena was alive and taken to the hospital where she was pronounced dead from a heart attack.

What television News reported that Erdogan had requested for asylum in the United States but changed his mind after the coup failed? Raises a question that the coup failed because or moles?.

# Murder Incorporated

From the CIA Black Ops by the Oklahoma City Bomber, Timothy McVeigh who wrote his sister that he and nine other men at Ft. Bragg, North Carolina were selected to do drug trafficking to finance covert operations and do government sanctioned assassinations. Wikileaks posted emails that Secretary of State, Hillary Clinton between members of her presidential campaign that were over the supplying shoulder fired stinger rockets to the Ansar al Sharia Libyan rebels without Congressional approval that would up in Afghanistan.

US Ambassador Christopher Stevens in Tripoli was told to supply the Libyan rebels with US made Stinger missiles to protect them against the Libyan Air Force of Muammar Gaddafi - without permission of US Congress.

So when a Chinook helicopter in Afghanistan was hit by a stinger missile on July 25th, 2012, but failed to explode because the Taliban who fired it failed to arm the missile, it lodged in the helicopter fuselage. After landing, an ordinance team recovered the stinger and traced the serial number to a cache of stinger missiles kept in Qatar by the CIA.

On October 16, 2016, *WhatDoesItMean.com*, by Sorcha Faal published on the internet that Hillary Clinton transferred 1.8 billion from the Clinton Foundation to the Qatar Central Bank *and* that the Clinton Foundation was a drug laundering operation.

This was bogus news but Megan Kelly broadcast it and got in trouble. Source of this bogus news ranged from Macedonian teens making fake news on the internet to get money from Google clicks to Google itself.

Mrs. Clinton sent Ambassador Andrew Steven *post haste* to Benghazi to get back the rest of the stingers from the rebels but Anser al Sharia attacked the Consulate and set it on fire and Stevens died from smoke inhalation. When Republican Congressman Trey Gowdy of the US House Select Committee on Benghazi asked Mrs. Clinton, "Why wasn't more protection given him?" Mrs. Clinton replied, "Well, it doesn't matter, they/re all dead now."

General Petraeus did not authorize the stingers sent to the rebels because these could be used on commercial airplanes. After President Obama pressured General Petraeus to testify that he did OK the shipment of arms to the Libyan rebels, Petraeus refused and was removed from his post for having a love affair with a journalist which compromised his security clearances.

The FBI was called to investigate Mrs. Clinton's use of a private email server which was used to hide everything from Congress. The Democrats in Congress said the investigation was an attempt to destroy her Presidential Campaign. Mrs. Clinton then began deleting emails, but it was too late a WikiLeaks had the evidence. Two weeks after the Benghazi attack, Mr. Obama told the UN that the attack was because of a u tube video. But the Taliban knew the inside story and pressured Mr. Obama to release of 5 Taliban Generals, which he did. But the news story was that they were release because the Taliban threatened to execute Sgt. Bergdahl who had been held captive by the Taliban for three years.

In November 2015, ten rebel Turkmen fighting Bashar Al Assad on the Syrian-Turkish border shot down a Russian SU 24 jet. Mr. Erdogan used NATO Article 5 Rules of Engagement to say the Russians had entered Turkey's air space. Then Turkish 0Prime Minister Ahmet Davutoglu said it was he who gave the order to shoot down the SU 24 Russian plane. After that provocation the Russians began listening to conversations between Erdogan and his son Bilal, who got permits to export ISIS stolen oil to Israel from his brother-in-law, Minister of Energy. Israel was only foreign buyer of ISIS oil.

A synthetic drug called *captagon* is used by ISIS fighters. Under its effect, ISIS fighter feel no pain from wounds and feel invincible. It is addictive, easy to make, and cheap to produce. Sold in Lebanon and the Syrian Black Market it is said that Saudi Arabia purchases 1/3 of the world's production but how that figure is arrived at, no source is given.

Muammar Gaddafi was accused by Israel and the CIA of supplying arms to the Sandinistas of Nicaragua. Not true. It was General Omar Torrijos, President of Panama, who supplied the Sandinistas with weapons and he asked Col. Gaddafi to give a list of serial numbers of weapons to hide his giving weapons to the Sandinistas. As Gaddafi had an Embassy in Panama he was anxious to establish embassies all over Latin America.

Israel supplied the Contras with weapons as well as Lt. Col Ollie North who used sales of cocaine from the Medillin Cartel to finance the war. After Colonel Noriega told the CIA that it was Torrijos who gave the Sandinistas weapons, the CIA put a bomb disguised as a radio in Torrijos' helicopter to kill him and Tony Noreiga made himself president and became best friends with Mike Harrari who planned the 9/11 attack to be blamed on Muslims and turn the whole world against Muslims. At that time foreign banks that came to Panama laundered drug money for the US and Israel that eventually tdeveloped into off-shore banking.

The body of White House Legal Counsel Vincent Foster was discovered in a small park off the Potomac Parkway two hours after Louis Freeh was named head of the FBI. Foster was killed hours before Freeh was named head of the FBI and called Vincent's murder a *suicide*. Freeh's FBI seized the letters the Oklahoma City bomber, Timothy McVeigh wrote his sister that he an 9 other men undergoing Special Forces Training at Ft. Bragg were hired by the CIA to do government sanctioned assassinations and drug tracking (to finance covert operations). Freeh also covered up Attorney General Janet Reno's order to burn the Dravidian Compound at Waco that killed 78 men, women, and children, and there was the shooting by an FBI agent of the wife of a man in Idaho who was resisting arrest and the shooting of an armed FBI agent who was armed and came onto the Wounded Knee Memorial on Sioux land.

Freeh resigned from the FBI a five months before 911. His replacement took over as the head of the FBI one day before 9/11

## Under Muammar Gadhafi

1. Electricity was free for Libyans.
2. Gasoline was US$0.14 a liter (US$ 0.46 a gallon).
3. 40 loaves of bread were US$ 1.15.
4. All banks were owned by the government and there was  no interest on loans.
5. The ownership of a home was a right.
6. All newlyweds were given US$ 50,000 toward the purchase of a house or an apartment.
7. A mother who gave birth got US$ 5,000.
8. Ant Libyan wanting to go into farming was given land,  a house, seed, equipment, and livestock free.

9. The world's largest irrigation project was the Great Manmade River that made water available everywhere.
10. Anyone not finding the education or medical treatment in Libya they needed and given an allowance of US$ 2,300 monthly for living accommodation plus a car allowance.
12. Fifty percent of a new car purchase was subsidied.
13. Education and all medical treatment was free.
14. 25% of all adult Libyans had a college degree.
15. If a Libyan was unable find work in his or her profession the government would pay him or her a commensurate salary of a professional until work was found.
16. A portion of oil sales was deposited every month is the bank account of all Libyan ctizens.
17. There was no external debt but there was US$ 350 billion in American bank account which is now frozen globally.


Julien Paul Assange is an Australian computer expert who founded WikiLeaks in 2006. In 2010, Bradley (Chelsea) Manning a private first class with top computer capabilities, copied classified US military documents to Assange's Wikileaks internet. The files exposed wars crimes committed the US Armed Forces. Both the invasions and wars of Iraq and Afghanistan Wars initiated by President George W. Bush were based on cooked CIA reports that Saddam Hussein had Weapons of Mass Destruction.

Julian Assange and President Donald Trump were on the short list pf *Time magazine "Person of the Year"* but the *Me Too* movement of women coming forth to complain about sexual harassment and groping was selected while President Trump was runner-up even though he has been accused of both.

In March 2003, British Prime Minister Tony Blair joined British Forces with the armed forces of the United States and invaded Iraq even though the UN assembly voted that Britain and the

United States America should wait until the UN Commission in Iraq finished its investigation on the charges that Saddam Hussein had no weapons of mass destruction.

A British board of *Congress information they had, Saddam Hussein posed a threat to the West.* No such investigation was carried out by America that would cast a shadow of guilt on the decisions by George W. Bush and US Congress committed a war crime by destroying the infrastructure of Iraq and causing the deaths of hundreds of thousands Iraqi civilians.

Private First Manning exposed war crimes committed by the military forces of the United States in Iraq and Afghanistan, the US Government imprisoned him and began following Assange disclosed a few of the war crimes that gained him celebrity status from anti-war groups all around the world. The American agents following Assange in Sweden learned of two girls who had sexual relations with him. They said Assange did not use a contraceptive thereby exposing them to potential AIDS infection and their charges were escalated to rape.

These rape charges were forwarded to England where Assange was welcomed by British anti-war groups. They offered their assistance to Mr. Assange and one of the British lawyers said that if Assange went to Sweden he could be easily extradited to the US as was Victor Bout who was extradited to the US from Thailand over alleged illegal arms sales in Latin American nations. Victor Bout has information on aid hehow the Pentagon obtained the Granit missile used on the Pentagon, September 11, 2001 and his stay in the Ecuadoran Embassy has lasted more than five years but has not silenced hm.

Julian Assange surrendered to the British police December 7, 2010 and was held in solitary and incommunicado for ten days. He was released on bail and exhausting all legal options in Britain, requested asylum in the Ecuadoran Embassy where the President of Ecuador had managed to stop a US campaign to overthrow him. In 1978, President Roldos, his wife and members of his government were killed in a helicopter crash in a rain storm in the Ecuadoran jungle. The natives of the area said was an explosion aboard the helicopter which questions if the CIA was involved as Roldos empathized with the Sandinistas. John Dimitri Negroponte held a minor position in the US Embassy in Ecuador at that time and then became US Ambassador to Honduras that was used as base for control of Guatemala, El Salvador, and Nicaragua where he covered up crimes of genocide, murder, disappearances, and drug trafficking that State Department.

Edward Snowden was the next to come forward with NSA documents that have revealed that British and American intelligence and the Mossad cooperated to create the Islamic State of Iraq and Syria (ISIS) and ISIL. Snowden showed that the intelligence services of the three countries have worked to create a terrorist organization to attract extremists from all over the world to one place, using a strategy called the *Hornet's Nest*. The NSA documents Edward Snowden copied refers to recent implementation of the *Hornet's Nest* by creating religious and Islamic slogans to be repeated again and again. These reports have been passed on to Israel's 8400 unit, to the Mossad, CIA, and MI6 to create an extremist group like ISIS. NSA documents released by Snowden state the *only solution* for the protection of the Jewish state *is to create an enemy near its borders* and reveal that ISIS leader and *cleric* Abu Bakr Al Baghdadi had intensive military training for a year by the Mossad, with courses in the Koran and speech making.

A Russian oriental studies expert, Vyacheslav Matuzov, has stated that the leader of the Islamic State of Iraq and the Levant (ISIL), Abu Bakr Al-Baghdadi, is in constant contact with the US Central Intelligence Agency (CIA). Al Baghdadi was a detainee in

Bucca prison in 2005 where he collaborated with the CIA," Matuzov told *Voice of Russia* radio the US does not use drones against ISIL as its commanders are US agents.

ISIL is a Takfiri extremist insurgency group that fought against the US occupation of Iraq in 2006 and joined with Syrian rebels in 2012 that are supplied and financed by the US. It is responsible for mass murders and destruction across Iraq and Syria. Israel claims that Saudi Arabia and the Emirates who support ISIS. However, a Kurdish Peshmerga (anti-ISIS) unit over-ran an ISIS stronghold and discovered Israeli military equipment and packed Israeli food in an Iraqi house in December 2012, but they have refrained from revealing more.

Earlier reports revealed that Israeli hospitals have treated the wounded ISIL militants fighting in Syria. Israeli Prime Minister Benyamin Netanyahu made a visit to a field hospital established by the Israeli authorities on the occupied Syrian territories that treat rebels fighting against the Assad government in Damascus. But ISIL headquarters are in the Kurdish cities of Mosul and Kirkuk.

In May of 2014, some 283 ISIL combatants were treated for wounds in the Israeli Zif Hospital in Safed. Hundreds of others have been treated at other Israeli hospitals after being wounded by the Syrian army that chased after them.

Zionists claim that the Protocols of the Elders of Zion is untrue and slanderous. But the best way to prove it is false and libelous is to stop acting like the *Protocols of Zionism* depicts Jews.

British intelligence has a history of duplicity from the mid 1800's as the CIA which was created after WW II. But as so many top CIA and Pentagon officials are Jews which is the same with British intelligence and the British military, the crimes of Israel have been documented but none have ever been tried in an international court as Britain and the United States use their veto power on the UN Security Council and elsewhere.

Every year since Israel's war of Independence in 1948, the US has used its veto vote in the UN to block all condemnation of Israel's war crimes, crimes against humanity, and Israel's Human Rights violations brought against Israel by Palestine and Israel's neighbors.

On June 11, 1967, 3 Israeli planes attacked the USS Liberty, an electronic spy ship that was spying on Egypt for Israel. These planes given to Israel by the United States for Israel's defense and without warming raked the ship's bridge at 2 pm with machine gun fire killing six American sailors and destroying all radio equipment. After 1/2 hour, 3 other planes replaced the first three. One plane napalmed the Liberty and five Israel torpedo boats appear and send five torpedoes at the ship. Only one hit and it blasted a 34 foot (10 meters) hole in the boiler room killing 28 more Americans. The ship's radio technician put together a radio to send out an SOS that reached the USS Saratoga did America know its greatest friend of the Middle East and only Democracy of the Middle East was capable of sinking and killing Americans to blame on Egypt.

The first attack on the USS Liberty was to destroy all the radios and severe the cables to the antennae of the ship so it could not call for help. The sailor who crawled out on deck to reconnect the radio cable to the antennae was wounded by rocket shrapnel but survived. The radioman found all five radio bands were jammed by interference from Tel Aviv. The only time the transmission overcame the jamming was when the Israeli planes strafed and sent rockets into the ship.

On the USS Saratoga, the calls for from the Liberty were put through the loudspeakers and all personnel could hear the machine guns firing on the Liberty. The Israeli attack continued for another 25 minutes when helicopters will with Israeli commandos in full combat gear came flew from Gaza and around the Liberty. Small arms were passed out with orders, "prepare to repel invaders". As Israeli helicopter pilots saw ships from the 6th Fleet appear on the horizon, they flew back to Gaza.

It is well known that President Johnson, as the as Commander in Chief of All American Armed Force ordered the planes back to the Saratoga and let the men of the USS Liberty die. Johnson the announced he would not to attempt a second election run for the White House.

Secretary of Defense McNamara sent Admiral Isaac Shelby Kidd Jr. to the USS Saratoga to order the wounded men being treated in the ships hospital that "Israel is to be exonerated of all blame. It was an accident and if any of you talk about this incident you will be subject to fines, prison, or *worse*."

One of the hospitalized men said he felt fear when Isaac Shelby Kidd said *worse* as he knew what *worse* meant. For years the men of the USS Liberty kept their silence until after they retired. They have created a USS Liberty memorial web site. A survivor of Israel's attack said, "If it was an *accident* it was the best planned accident I ever knew of."

A pictures of the severely wounded of the USS Liberty in the hospital of the USS Saratoga is in thr photo section of this book. Admiral Issac Shelby Kidd was flown to the Saratoga to tell than that if any of them talked about Israel killing 34 men and wounding 174, they would be fined, imprisoned, or worse," Israel tried to sink the ship and kill all of the men to blame on Egypt so Americans would be screaming for Israel punish Egypt and annex the Sinai, West Bank, Gaza, and Golan Heights and stay on the Suez Canal. Treason was committed on June 11, 1967 by Americans in the White House, Pentagon, NSA, and CIA just like 9/11, and the murder of President Kennedy.

Only a bit of the dialogue of Amy Sweeny and Betty Ong, the two flight attendants whose conversations with American Airlines offices by phone are recorded.  They tell how the plane was hijacked by Middle Eastern looking men who stabbed two flight attendants, broke into the cockpit and seized control of the plane.  The suicide pilot was supposed to be Mohammed Atta but his airline ticket to come to America to learn how to fly planes was paid for by the Mossad.

A video showing the plane hitting the South Tower making a 2 ½ Gravity turn then crashing into the Tower making a gigantic fireball is a decoy.  One airline pilot on viewing the video said the 767 Boeing that crashed into the South Tower looked like a refueling tanker given to Israel, but it was a decoy video to mislead *as no plane hit the South Tower.*

According to Wikipedia, there were four hijacked flights on September 11, 2001.  United Airline Flight 175 to the North Tower\ American Airlines to the South Tower, American Airlines Flight 77 to the Pentagon and United Airlines Flight 83 that came apart for over a mile of bodies, seats, and wreckage of the disintegrating plane.  The truth is that Flight 83 pilots strayed off course and came under suspicion that the plane was hijacked.  Attempts to contact the pilots alerted the Maryland Air National Guard.  One Maryland National Guard pilot said Flight 83 was not shot down, but Secretary of Defense Donald Rumsfeld said the next day that the plane was shot down.  That information is no longer on the internet.

The author has had four attempts on his life and a successful.  Seven years later he went to the offices of the doctors that saved him to tell them his heart was normal again but all were now barred from medicine but the killer still practices medicine in Houston, Texas  Interesting, no?

249

The author learned in February 2016 that Homeland Security visited his ex-wife to whom he hasn't spoken to for fifty years to get her to say that he knew some Turks when they were married. They hammered on her until she said yes. Crazy because the author didn't know any Turks until he met one in China in 2002. That Turk was a Jew whose ancestor came from Spain to Turkey in 1492 when Sultan Mohammed sent ships to bring 30,000 to Turkey to worship as they pleased. His ancestor had to leave Spain if they didn't want to convert to Christianity but he hid his Jewishness and said he was Muslim, married a lovely Chinese Muslim girl but he was fully Jew as were the Fatih Koleji school management that Erdogan's former friend, Gulen invested in and built all over Turkey and now employs in Muslim schools across the United States. Erdogan tried to blame Gulen for the attempted coup but Gulen, who lives in Pennsylvania had nothing to do with the coup.

So why was Homeland Security hammering on his ex-wife? Well it seems they were trying to say the author was recruiting for ISIS. Crazy as ISIS was created by Israel's Likud Party that sent Mossad agents into detention centers in Iraq where they tortured Muslims more than was done at Guantanamo by Marines.

Please note that Homeland Security uses information from anti-personality disordered persons as verifiable information.

President Barak Obama said in January 2016 that Muslims killed 3,000 Americans on September 11, 2001. No, Mr. president, it was Mossad Colonel Mike Harrari, Israeli PM Areal Sharon, President George W Bush, Vice President Dick Cheney, and members of the Armed forces who did 9/11. Dimitri Khalezov told the truth and the proof is in the photos of the deep craters made in the bedrock by mini-nukes

'Treachery" has some of the author's own experiences and information from people with high standards of morality. The murder of John F .Kennedy was carried out by people who took an oath to protect the United State and the president. The Warren Commission was composed of eight men. House Minority leader, Gerald was informed 16 days before the assassination and kept quiet and he became president. The Warren Report full of lies and crooked evidence put the blame on Lee Harvey Oswald.

Simon Wolf of the Confederate spy ring and head of B'nai B'rith manipulated John Wilkes Booth to kill Lincoln so as to deliberately impoverish the South, re-enslave the newly freed slaves by instituting tenant farming and to create racism that destroyed the reputation and moral values of America.

President Woodrow Wilson was manipulated into WW I that was created by the bankers who formed the Federal Reserve Bank that was used to finance WW I with American money in loans that were never repaid.

World War II was started by Winston Churchill using the combined armies of the League of Nations and President Franklin D. Roosevelt Britain had to *save Great Britaialestine.* But he called making the world safe for democracy when more than three quarters of the world was under empires or Communist dictatorship.

The United States, Britain, and Israel employ assassins, spin doctors, and liars to kills, confuse, and blame others for corruption.

The Supreme Court has failed to protect the Constitutional *rights* and *human rights* of Americans since September 11, 2001. The *Patriot Act* sanctions unlimited detention, government can withhold evidence and charges if the government so pleases.

Assassination units commit murder municipal, county, State, and Federal law enforcement have labeled suicide.

Senator Church's Commission began an investigation after May 9, 1975 into the activities of 14 intelligence agencies that covered spying by the US Military on the American civilian population, the opening of US mail by the CIA without a court order, and President Eisenhower called on the Mafia to assassinate Fidel Castro in a plot devised by Allen Dulles, head of the CIA.

A video on u tube showed Senator Barry Goldwater smiling as he passed the *assassination gun* that would fire a needle shaped charge into a victim that would dissolve in the victim's body, inducing a heart attack or releases cancer viruses.

Senator Goldwater was the Republican presidential candidate who ran against Lyndon Baines Johnson who covered up Israeli's attack on the USS Liberty three years after JFK was assassinated.

Israel is rated as the Number One nation in the business of assassinations with the United /states, Great Britain, Russia, and France next. When the murder is more public, a scapegoat will be found by a government agency and other agencies will join in the cover up these act of murd operation like 9/11.

The False Flag/Black Operation of 9/11 went beyond 9/11 to kill millions, and destroy Iraq, Afghanistan, Libya, and Lebanon and attempting to overthrow the democratically elected government of Bashar Assad. The US has used the tactic of debasing currencies, creating inflation, embargoing trade, and taking control of the mineral resources of many counties. The exchange of weapons for drugs has increased availability corruption that has affected the minds and morality of people in the United States, Britain, Israel, and nations under the control of these three nations..

3. of Investigation or government agency to join in covering up these act of murder such as was done officials in the government also contributing to this operation like 9/11.
4. The False Flag/Black Ops of 9/11 have gone beyond 9/11 kill millions, and destroy Iraq, Afghanistan, Syria, Libya, and Lebanon and at the same time debasing currencies,

creating inflation, and taking control of the mineral resources and increasing the wealth of those involved. A side effect is the consumption of drugs affecting the minds and morality of the people of the United States, Britain, and Israel and allied nations.

There are studies of the prehistoric earthquakes, eruptions, tsunamis, climatic changes, atmospheric changes and shifts of the magnetic poles that are close to reemerging again such as the Great Rift Valley that begins at the Sea of Galilee, goes through the Jordan River, Dead Sea, the Red Sea, through the Horn of Africa, to Lake Malawi and Lake Tanganyika of East Africa all for a distance of 6,000 kilometers.

In the United States there is the New Madrid fault the that unleashed and earthquake in 1815 that had the mile wide Mississippi River going backward so that when it returned to flow south again in a rush it created the Reel Lakes of Southern Illinois. Land was described by the people at time as rising and falling like waves in the ocean as the waves traveled away from the epicenter of the New Madrid, Missouri epicenter.

New York City's Manhattan Island was a huge volcano that was removed by the Ice Age and the Palisades on the Hudson is a *sill* which is molten extruded into a fault line that has been exposed by weathering or erosion.

Yellowstone Park has a 30 to 35 mile wide caldera with a mega magma below it that if unleashed could destroy and poison the land around if for hundreds of miles. Methane gas is leaking in the mountains north of Los Angeles, and there are volcanos of Alaska and Hawaii to be considered. Where Australia and South America were once joined together, they are now thousands of miles apart. And peaceful Switzerland was formed from massive volcanism. Countries around the world show that they have been developed from massive movements of the Earth's crust of within the Earth.

The possibility of climate control of the Earth and potential to harness and redirect the energy in the Earth's crust and atmospheric is a possibility. Nikolai Tesla said there was free energy that would make the life of humans easier and more comfortable. While the wars of the 20th Century profited those who control and exploit natural gas, oil, thorium, and nuclear power.

Robert McNamara ordered a report on the Viet Nam war that Daniel Ellsberg copied the report, which was called *the Pentagon papers,* and he gave it to the *New York Times* that published it. The result was that the Pentagon Papers showed that President Johnson had lied to Congress and the American people and Johnson declined to run again for a second term in the White House. Ellsberg called Edward Snowden a Patriot for exposing NSA spying on Americans, Snowden has not responded as far as the author knows.

The highly decorated Marine Major General Smedley Butler exposied the attempted coup against President Roosevelt in 1933 and testified before the US House of Representatives Special Commission on Un-American Activities. E wrote abook many books are coming out now to herald book about it and *The New York Times* called it a gigantic hoax. But it was the truth.

Gerald C. MacGuire was the go-between the plotters and General Butler. Butler was promised an army of 500,000 Veterns of Foreign (VFW) from WW I to march on Washington. The VFW was later used by Roosevelt to popularize his WW II. The leaders of large Corporations such as Heinz, Colgate, Birds Eye, General Motors, and banks in the Federal Reserve, Felix Warburg, J.P. Morgan, Henry Ford, Allen and John Foster Dulles, George

Herbert Walker, and Prescott Bush (both grandfathers of George W, Bush) were in the plot to overthrow Roosevelt..

Major General Smedley Butler was respected by the Veterans of Foreign Wars for refusing to clear the tent city the veterans had created in the Mall from Capitol Hill to Lincoln Memorial. President Herbert Hoover ordered General MacArthur clear them out and he used the US Calvary. Congress covered up this treason. But the wealthy were right. Roosevelt was a liar and his plans to bring the US out of the Great Depression raised prices, unemployment and the Federal Reserve foreclosed on millions and thousands committed suicide.

The main focus of this book is the murder of 3,000 Americans and the theft of 2,3 trillion dollars and 800 million in gold on September 11, 2001 and aided by Americans who were responsible for the safety and security of the Unite States. To incriminate Muslims and Islam has been blamed on the US, not Israel.

*Operation Hammer* was a scheme by George H.W. Bush to grab for the gas and petroleum assets of the Soviet Union. The sanctions to destroy the Russian economy and keep Vladimir Putin out of the Middle East that has the Russians retreating from Syria due to the drain on the Russian economy. All the problems about the Crimea and Ukraine appear to be diversions to keep the investigation of 9/11 from focusing on American traitor aiding the Israel's 9/11.

WikiLeaks has shown Israel and Britain conspired in creating the Hornet's Nest, also known as *Plan Britannique* that became ISIS, Al Nusra, Syrian rebels, and Turkish rebels fighting Syria.

In 1953, President Dwight Eisenhower appointed the Dulles Brothers, Allen as head of the CIA and John Foster Dulles as Secretary of State respectively. They formulated foreign policy. The so-called non-profit think tanks. The Rockefeller Foundation which founded May 14, 1913, by John D. Rockefeller Sr, which was a tax dodge for Standard Oil. Its first president was Frederick Taylor Gates, and the stated goal was *to improve the well-being of humanity throughout the world.*

Then came the Brookings Institution, founded in 1916 and now headed by Strobe Talbot who was made Under Secretary of State by his Oxford University roommate Bill Clinton. Strobe' support of Boris Yeltsin as the first president of Russia was being undermined General Pavel Grachev who financed Serbia's war in Bosnia to hold the Yugoslavia together to menace the right flank of NATO once the Russian military began to remake he Soviet Union and move back into the Ukraine, Poland, and the Balkans.

John F, Kennedy was upset when the CIA murdered the Diem brothers in Viet Nam, but Prime Minister of Israeli, David ben Gurion, had no qualms about ordering the murder of John F. Kennedy. He was ready to implicate Cuba and Russia as responsible for JFK's murder so that Americans would push for retaliation which the Pentagon wanted to do.

Henry L. Stimson, a graduate of Yale, member of the Skull and Bones secret society, and a Republican was appointed Secretary of Defense under Franklin D. Roosevelt and he pushed for war with Japan and Germany. In 1943, he ordered the creation of the biological warfare laboratory at Ft. Dietrich, in Maryland and work began immediately on developing an anthrax bomb. Churchill told Hitler that if he didn't stop bombing England with missiles he would get anthrax bombs from Roosevelt and use them on four German cities. Hitler stopped sending the V2 missiles into England.

In 1969. President Richard Nixon ordered the biological warfare laboratories closed, but they were kept open to develop Agent Orange and drugs used in the MKUltra mind control program that experimented on hospitalized people without their

256

knowledge or permission. Helms destroyed the MKUltra records and convicted of lying to Congress, a felony, but never sent to prison. Helms was especially proud of his conviction.

The author suggests that the first step for peace and reduce world tensions would be full international inspection of all nuclear technology of all nations that have nuclear weapons and nuclear power. Any nation refusing to conform to this order shall lose the rights and membership in the United Nations with sanctions applied until agreeing to cooperate fully.

Second step is that all nations must accept the authority of an International Human Right Commission as all nations must adhere to a code of conduct that forbids the detention or imprisonment of anyone whether citizen or foreign born, torture, use surgical procedure against a person's will, or mete out humiliating treatment, sexual harassment.

All prisons, concentration camps, or camps of confinement must be open to inspection where Red Cross or Red Crescent can enter and talk to any inmate at any time. All authority figures who have committed atrocities, human rights abuses must be turned over to the international tribunals and any nation that refuses will be excluded from the United Nations. The United Nations must improve as its record has been dismal if not criminal.

All nations whose governments permit assassinations, plan False Flag/Black Ops, or torture, shall disband these groups immediately or be branded a terrorist nation and treated as such.

The author deliberated a year before including the danger of an H bomb dropped on Yellowstone's 30 mile by 35 mile caldera. I was only after he read Christopher Brenman's article of April 1, 2015, 13:03, in the *Mail On Line,* published by *Dailymail.com* about retired Kremlin military analyst, Konstantin Sivkov's article, *Nuclear Special Forces* did he include it *Sivkov* wrote that a powerful atom bomb on Yellowstone's caldera would make it erupt and clouds of gas and volcanic dust would keep sunlight from reaching more than half the United States, spread trillions of tons of ash, and destroy crops and create famine. Sivkov also suggested placing atomic bombs above the ocean floor off the coasts of the United States to create tsunamis that would affect 80% of Americanpopulation that live within a 100 miles of America's coastal litoral. **He said the tsunamis would also damage some of America's European Allies.**

Konstantin Sivkov wrote his that *Nuclear Special Force* would serve as an asymmetric threat to the border of Russia. Sivkov is president of Geopolitical Problems and wrote *One Day Without America, The Day After Tomorrow, and Apocolypse Simply and Cheaply* for *Vorkantankte* (2,000 subscribers).

Russian television presenter on the States television, Dimitri Kiselyov said Russia is the only country capable of turning the United States into radioactive dust. While many Americans are cheering the Russians taking down the so-called Islamic State (Desh) in northern Syria and Iraq, many Russians are unhappy of the embargoes and blocking international currency transfers. Putin warns that his call for nuclear de-escalation have been ignored by American counterparts.


The author points to the huge Rift Valley that runs from the Sea of Galilee, down the Jordan River, trough the Dead Sea, Red Sea, Lake Uganda to Zimbabwe and passes more than one mile below Israel's Dimona nuclear bomb facility and storage puts this fragile region as the Apex of the Earth's extinction.

Each of the 92 nuclear warheads at Incerlik Air Base in Turkey is only rated at 100 times more powerful than the atomic bomb dropped on Hiroshima on Aug, 6, 1945. But the Russian Czar Bomb tested on Nova Zemlya in 1960 was 1,570 times more powerful than Hiroshima bomb. President Putin wants discuss what should be done with 34,000 tons of highly fissionable Plutonium has gone unheeded by Washington.

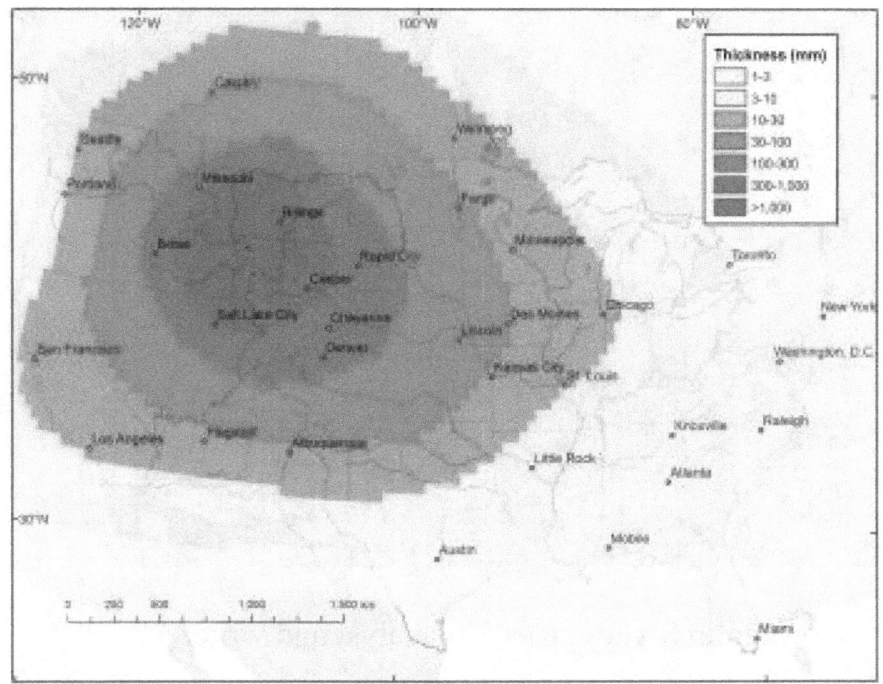

Map by the Geological Survey of the United States showing amount of volcanic ash expected from Yellowstone.

The author used the measurements of French scientists is assessing the vaporization of the bedrock by the mini-nukes placed in the basement of the twin towers to say an H bomb on the caldera of Yellowstone would cause it to erupt.

Russia's Putin has threatened to use nuclear weapons but did not say on whom he would direct on the US, Israel, or NATO. The United States, EU, and UN have put embargoes and financial sanctions on Russia which have hurt Russian exports by depriving Russia from international currency payments through Mastercard or Visa, both controlled by the Federal Reserve forcing Vladimir Putin to resort to off-shore banking like Panama's Fonseca Mossac previously mentioned..

Konstantin Sivkov, Russian analyst and writer

'

Israel planned a nuclear attack on Iran's nuclear energy program in 2013. It did not come off because two generals said there was no information on how far Iran had progressed in its atomic power program. A nuclear bomb of the facility might make more damage that would extend beyond the area to be destroyed.

Iran's elected leader was the leader of the non-aligned nations movement in the UN that is against making nuclear weapons. Mr. Obama signed a nuclear deal with Iran allowing it to proceed in its program for nuclear power but applied more sanctions to Iran that have erupted into violent demonstrations that Mr. Trump has expressed his sympathy for.

More than a year has passed since Mr. Obama said Israel must comply to the international inspection and control of its nuclear weapons facility and storage which Israel has ignored as did David ben Gurion who ignored president Kennedy and gave the order to kill president Kennedy. That was carried out by Jews in the CIA that manipulated people in the Pentagon, Secret Service, and FBI.

The author documented the fragility of the Great Rift Valley and will state that an H bomb detonated on Israel's Dead Sea salt deposits, one mile below sea leaven, will produce vaporized radioactive sodium, thousands of times more dangerous than the Chernobyl nuclear accident and generate even more heat than the mini-nukes placed on the basement floors of the twin towers of the WTC. Could the Great Rift fault line running from the Sea of Galilee to Mozambique in Africa crack open and reduce the 2.2 million atmospheres of pressure on the liquid iron of the upper mantle of the Earth's core that holds the Earth's atmosphere to the Earth destroy the Earth like the planet between Mars and Jupiter?

After Edward Snowden worked for Booze, Allen, and Hamilton, whose CEO was Dov Zakheim who stole 2.3 trillion dollars from the Pentagon budget. While Mr. Trump says Mr. Snowden is a traitor for exposing that NAS was spying on all Americans, but **he says nothing about Dov Zakheim's theft of 2.3 trillion dollars, Dov most certainly knew about 911 as well as Larry Silverstein who told the New York City Fire Marshall to "pull building 7 as there has been so much terrible loss of Life."**

This book supports Dr, Rainer Karlsch's document that shows Germany made the first atom bombs as America it is the nuclear centrifuge that made it possible. American physicists created the Cyclotron at Batavia, Illinois that is now used to study sub atomic particles like the huge atom splitter at CERN on the French-Swiss border and Lop Nor, Xingjian, China, The author documents that Mussolini was going to Berlin to be with Hitler when he launched a counter0attck to drive the Russians back to Moscow which came from a British soldier who was *captured by the German Army in North Africa, sent to a German POW camp in Italy from which he escaped and was led to Serbs brought to*

*Italy as slave labor for the German Army and who escaped to form a partisan group that was killing German soldiers for weapons and ammunition. It was they who captured Mussolini and asked London what to do. Wikepedia states as fact that that the Czar Bomb tested in 1960 in Nova Zemlya was 1,570 time more energy than the atom bomb dropped on Hiroshima in 1945.*

*This book documents torture, war crimes, murders, theft, lies, and corruption, with facts from Encyclopedia Britannica and New York Times Index, research by Kevin Ryan the insurance claims by Larry Silverstein, PANYHJ, and the information from persons who have witnessed the crimes. The author states all the information in this book as true without political or financial bias and has no financial support or political backing of any group or person.*

*While the Attorney General of Mr. Bush and Mr. Obama have hounded, blocked, hindered, burdened, and pushed for the arrest of Assange on slander of rape, Mr. Snowden is called a traitor who should be executed by Mr. Trump which shows he could not get a fair trial in the United States and the charges against him would have to be dismissed.       The CIA oversees Hollywood stopped a movie producer from making a movie of his story on Bosnia. Homeland Security tried to implicate the author with ISIS in Turkey, and the wedding ceremony in Guangzhou the night of 911 with the dancing men on the hotel stage who were not videotaped by the author's boos but who videotaped the 71 year old author dancing for joy at the news of 911 and turned over to China's secret police, and State Department working to change rules to get the author out of Bosnia and other countries has been less than pleasant.*

The author supports John Lear's document that no plane hit either twin tower or the Pentagon and that David ben Gurion ordered the murder of President Kennedy and John Rich and James Jesus Angleton of the CIA carried it out. The sons of E. Howard Hunt told their their father confessed of his involvement was rebuffed by the CIA.

Joe Kennedy Jr. was killed in 1943, not at the end of WW II, Why was this kept secret until 2008? Joseph Kennedy Sr. was grooming Joe Jr. to run for the office of the President of the United States but so was Franklin Roosevelt Sr. grooming his son to be President of the United States. And Franklin Jr. was in the mother ship that controlled the demolition drone Joe Jr. was flying. Would Franklin D. Roosevelt Jr. have won an election for the presidency if his father had lived instead of dying, April 1, 1945?

Somehow those who run the political system chose to ignore Franklin Jr.. But the important information is that 20 tons of explosive in those drones never could make a dent in the steel reinforced concrete. A mini-nuke weighing a kilo of so will have one to three kilotons of explosive power and 6,500 C. that is in the range of plasma.

Russian nuclear intelligence officer, *Dimitri Khalezov, states mini-nukes were placed on the basement floors of both WTC twin towers. A photo of crater made by the mini-nukes that vaporized the bedrock on which the twin towers were built is the undeniable proof with silver colored steel and two* bubbles created at a temperature greater than 6,000 C.

The Greeks named the four elements as air, water, fire, and Earth. These are not elements but states of the elements determined by temperature. Voltaire brought Newton to the attention of French scientists who then devised the Celsius thermometer based on the temperature of water changing into vapor at sea level and the lengths, volume, and weight were then put into 10 million lengths (meters) from the equator to the North Pole passing through Paris. That has since been refined to the distance light travels and the speed of light has been given in both the old measurement of miles and kilometers, and temperature is

263

now also measured in Kelvin, in which zero is absolute zero. If you are reading a book, the molecules of atoms composing the paper are in constant motion. Atoms remain the same in size and do not expand but motion speeds up and therefore a hot air balloon will ascend at the amount of air in the balloon is lighter than the hot air inside that is pushing against the walls of the balloon and forcing air out.

A plume of hot vaporized bedrock came up through the center of the collapsing tower and rose to about 1,000 feet before it was blown toward the sea, while the dust from reinforced concrete and steel fall down as debris following Newton's law of gravity. That plume would be the vaporized bedrock that was displaced by the temperature of the mini-nuke. Rock is composed of around 50% oxygen, unlike steel which is carbon and iron and alloys without the lighter element of oxygen which can act as a catalyst. Therefore the melted silver colored steel is draped over the rugged floor of the bottom of the crater formed by the mini-nuke.

The steel reinforced concrete wall 0f the Pentagon was hit by a intelligent Russian missile designed to sink an aircraft carrier by traveling two meters above the surface of the ocean or land. Russian arms dealer Viktor Bout was extradited from Thailand and imprisoned in a prison in the United States to silence him as he knows how the Russian missile landed up in the hands of the US Naval Research lab, from which it was taken to kill the accountants on the first day of going through the account books of Rabbi Dov Zakheim to find where he hid 2.3 trillion dollars of the Pentagon budget that disappeared while under his control as Pentagon comptroller who President Bush personally appointed to that position.

To verify my estimates which I based the estimated 6,500 C vaporizing the bedrock under the twin tower was based on the measurement of the temperature of the liquid iron of the upper mantle of the Earth's core by French Scientists in 2015,  To get a verified answer, I went to the nuclear science faculty of the

University of Illinois to confirm or state what I used was accurate. At the nuclear lab where classes are held, my wife and I were escorted out of the building as not *being allowed to be there.* Absolutely ridiculous as there are classrooms for and this is a public university and one of eight universities I have attended.

I went to the library to find the head of nuclear science department but just missed him. I was then directed to the office of the assistant to the head of nuclear science. He listened to what I wanted and then said he was too busy. *The Fix* is in. No academician or politician will dispute the abysmal NIST Report on 911 for fear of losing his job or losing his seat in Congress. The Fix is like the Mafia and the Mob when arrested and charged with a crime pays money to fix it so the charges will be dismissed. If one does not comply he is either killed, bankrupted, or jobless.

Edward Snowden only copied classified documents of NSA spying on people of the United States. Pentagon Comptroller, Dov Zakheim stole 2.3 trillion dollars that would pay for all American credit card debt, student loan debt, and Medicare. Who is the biggest criminal? Are Julian Assange, Bradley Manning, Edward Snowden the biggest criminals, or is Rabbi Dov Zakheim of the Pentagon and Booze Allen, Hamilton the criminal?

What *Treachery* show is that there is an almost complete lack of integrity, ethics, and humanity in the top leadership of the United States, Israel, Turkey, and Great Britain while Mr. Putin accedes to Russian scientists, church leaders, and others in Russia, none of the leaders of the aforementioned countries have acted on their moe on their own and if there is any objection to their policies it is not voiced, such as the complete silence on the evidence 9/11 was not and could not have been done by these amateurs that could get past immigration, customs, and more. To believe it was possible for them to do so is like believing in fairies, Superman, Spiderman, Batman, is out there to protect Americans while the government plans more wars and an explanation that torture is need to find out what detainees plan to do next in the way of terrorism.

Russian electronic technicians have invented electronic weapons that do no kill but shut down the destroyer, USS Cook which is loaded electronically guided missiles and left the USS Cook stranded without even power to move. When President Trump ordered Tomahawk missiles to bomb a Syrian Air Base on allegations that Basher Assad used sarin gas on rebels, Mr. Putin remarked that he should have waited to verify that as sarin gas was given to the rebels by the US.

As the Tomahawks are electronically fired and guided, the latest news is that the Russians have disabled all of the Tomahawk missiles with this electronic technical weapon.

# Postscript

With President Trumps announcement: "9/11, cas closed," I will tell how I have been subjected to monitoring and spying by government agents and agencies, four attempts on my life and one where I was murdered in Houston, Texas, October 27, 2003, in a hospital. The time is perilously close to a nuclear war that is the result of Israel and its Israeli Lobby that has seized our country's financial assets, educational system, used out billions of economic aid without which Israel would have collapsed long ago. Below is one of three EKGs made in a heart clinic in Kuala Lumpur, Malaysia, in May 2004. Chinese medical exams show blood clots under my armpits and the left lower ventricle of my heart with is valve pulled backwards and going up and down as shown my my doctor in KL.

The mother smothered me with a plastic oxygen mask. I was in a deep sleep when is suddenly I awoke and my heart was pounding rapidly, I could not push away the murderer *dressed in surgical greens with a cap and mask on* I was tied to the gurney by the nurses to keep me from falling off  They left me alone giving my killer the opportunity to murder me.

I could see his eyes and high bridge nose between the green cap and surgical mask and as he pushed the mask hard on my face his eyes were intense.  He was not a Mossad as he would have given me muscle relaxant collapsing my lungs making it impossible to resuscitate me.

There are basically three Jewish sects: Orthadox, Reform, and Conservative.  The Zealots of Jesus time carried knives in which to kill non-Jews and drive Greeks, Philistines, Romans, and all non-Jews out of Palestine in the same way Zionists are treating Muslim and Christian Palestinians today.  These Jews adhere to the Talmuc which has no connection to Mosaic Law which Christian adhere to.

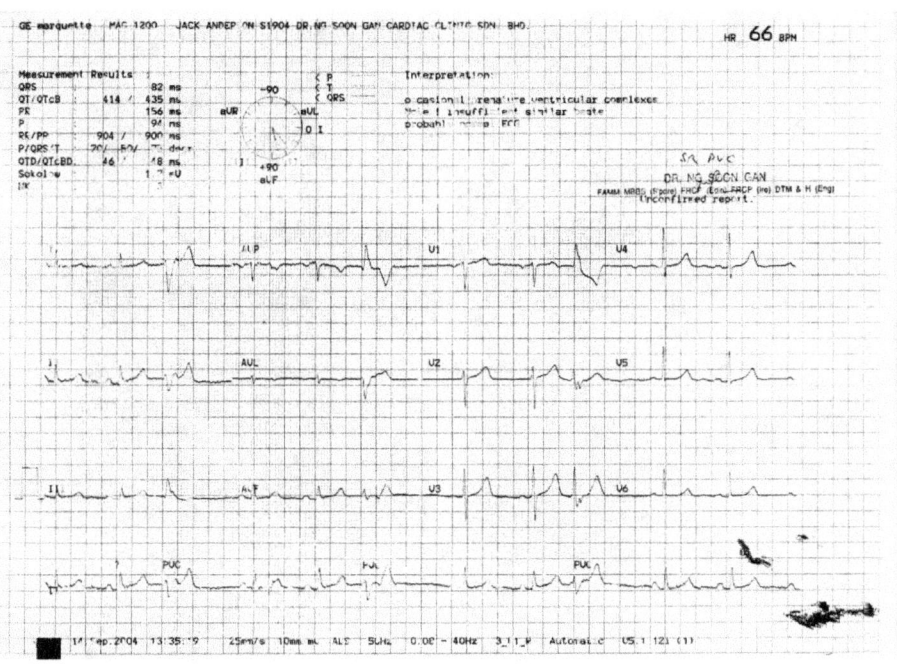

广东省肇庆市第一医院

姓名: Andy          ID:          性别: 男    年龄: 78   日期: 5-20-2008

病区: 特需          床号:          临床诊断: 体检          :

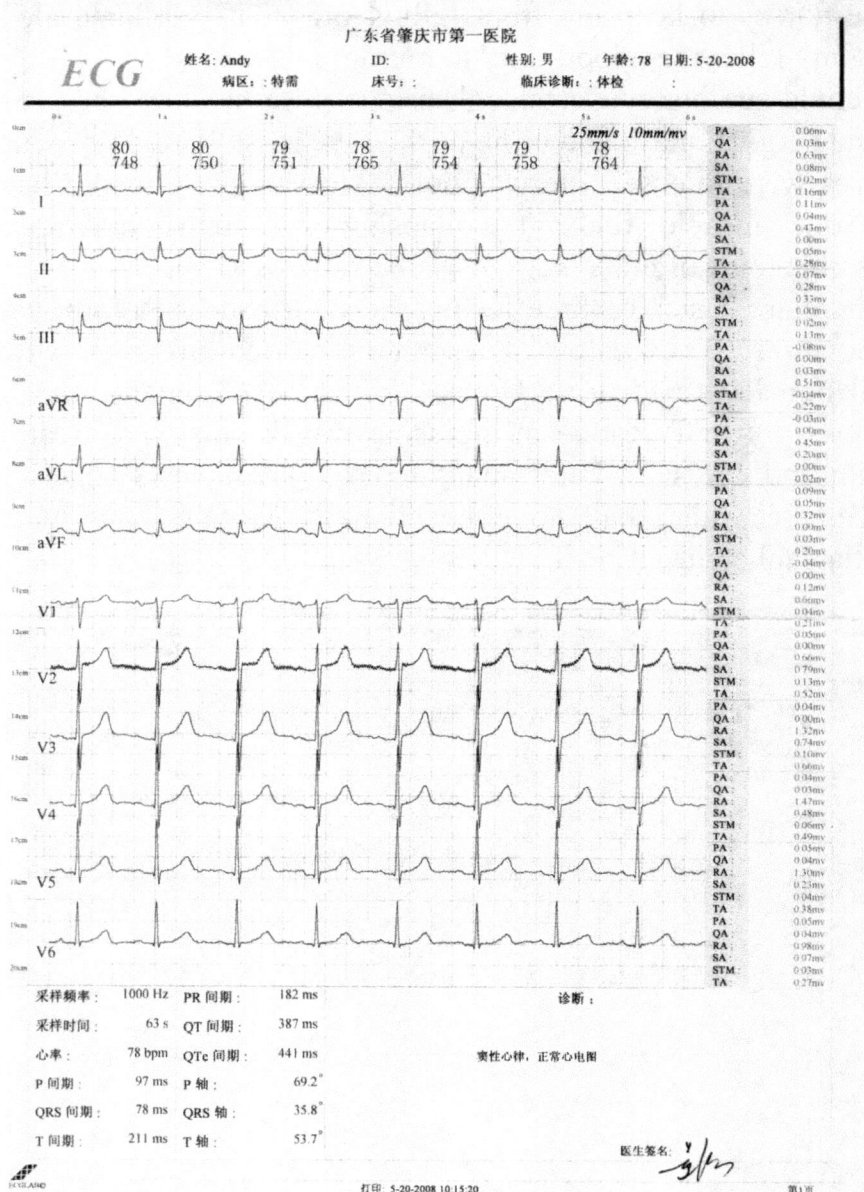

| PA: | 0.06mv |
|---|---|
| QA: | 0.03mv |
| RA: | 0.63mv |
| SA: | 0.08mv |
| STM: | 0.16mv |
| TA: | 0.11mv |
| PA: | 0.04mv |
| QA: | 0.43mv |
| RA: | 0.00mv |
| SA: | 0.05mv |
| STM: | 0.26mv |
| TA: | 0.28mv |
| PA: | 0.00mv |
| QA: | 0.13mv |
| RA: | -0.08mv |
| SA: | 0.43mv |
| STM: | 0.51mv |
| TA: | -0.04mv |
| PA: | -0.22mv |
| QA: | -0.03mv |
| RA: | 0.45mv |
| SA: | 0.20mv |
| STM: | 0.02mv |
| TA: | 0.05mv |
| PA: | 0.03mv |
| QA: | 0.20mv |
| RA: | -0.04mv |
| SA: | 0.12mv |
| STM: | 0.04mv |
| TA: | 0.21mv |
| PA: | 0.05mv |
| QA: | 0.00mv |
| RA: | 0.66mv |
| SA: | 0.79mv |
| STM: | 0.13mv |
| TA: | 0.52mv |
| PA: | 0.04mv |
| QA: | 0.00mv |
| RA: | 1.32mv |
| SA: | 0.74mv |
| STM: | 0.10mv |
| TA: | 0.66mv |
| PA: | 0.03mv |
| QA: | 1.47mv |
| RA: | 0.48mv |
| SA: | 0.06mv |
| STM: | 0.49mv |
| TA: | 0.04mv |
| PA: | 1.30mv |
| QA: | 0.23mv |
| RA: | 0.04mv |
| SA: | 0.38mv |
| STM: | 0.05mv |
| TA: | 0.98mv |
| PA: | 0.07mv |
| QA: | 0.03mv |
| RA: | 0.27mv |

25mm/s 10mm/mv

| | | | | |
|---|---|---|---|---|
| 采样频率: | 1000 Hz | PR 间期: | 182 ms | 诊断: |
| 采样时间: | 63 s | QT 间期: | 387 ms | |
| 心率: | 78 bpm | QTc 间期: | 441 ms | 窦性心律, 正常心电图 |
| P 间期: | 97 ms | P 轴: | 69.2° | |
| QRS 间期: | 78 ms | QRS 轴: | 35.8° | |
| T 间期: | 211 ms | T 轴: | 53.7° | |

医生签名:

打印: 5-20-2008 10:15:20                    第1页

nnoo

I am a descendant of James Patterson, whose father, Arthur Patterson was executed by Major Patrick Ferguson, s *renegade Scot,* the day before the biggest battle and greatest victory of the

269

American Revolution. James and his brothers William and Thomas were joined by 900 men of the militia army created by Col. Isaac Shelby to battle Ferguson. That victory came after General Gates lost 4,000 men at Camden, New Jersey and Lincoln lost 5,000 men and three signatories of the Declaration of Independence at Charleston, South Carolina.

James and three of his sons fought at the Rampart outside New Orleans in War of 1812. The British sent 5,000 Highlanders to charge the mainly French and American Scots. All were killed or wounded and the British surrendered. Another victory my ancestor was in and I am old and tired of persecution by my government.

Another ancestor was Augustine Herman who did the first map of Maryland and was made a 'Citizen of America' by the Maryland over one hundred years before the creation of the United States of America in 1791. And General Wood was a descendant through his daughter and he served on George Washington's staff.

I have been bugged, blocked, slandered, and defrauded by Jews, Mossad, JDL ADL in America and all the 10 countries I have lived in and many of the 50 countries I have traveled in.

Mr. Bush, Mr. Obama, and now Mr. Trump are covering up the murder of at least 3,000 Americans on September 11, 2001. I have been murdered, had four attempts on my life and I am fighting for my country and return of honor to my country that my ancestors fought and died to create a nation of freedom that has not always acted perfectly in relation to native Americans, Afro-Americans, or Asians.

I am angry over the persecution of my children and others who befriended me. Zionist Jews act like the Zealots who were so much trouble, the Romans dispersed the Jews around the Mediterranean. As for the Khazar Jews who have infiltrated our Supreme Court, Congress, White House, Pentagon, and Treasury, and ,secret societies, I would recommend taking away their pensions like was done to Ollie North who is Catholic but is an example of those who seek promotions and wealth by the positions they are entrusted with to America. I support the actions of men like Edward Snowden, Chelsea manning, Julian Assange, Jimmy Wales, and Christopher Boleyn for bringing truth into this world.

As for the evidence showing Israel's Mossod Colonel Mike Harrari boasting to his friend, Dimitri Khalezov that he was the mastermind of 9/11 and dying of all kinds of cancers, there was also Israel's top military commander, Ariel Sharon, who was Israel's Prime Minister at the time of 9/11, went into a coma and died six months later, there will be no welcome by St. Peter at the Golden Gates to Heaven. Well, it doesn't matters as neither believed in heaven or hell as they were Talmudist who rejected the Torah and the Ten Commandments given to Moses that has been used by Muslims, Christians, and Jews for more than 20 centuries,

One final note. The divorce between President Donald Trump and his first wife, Ivanka, was finalized March 24, 1991. It has been said that Mr. Trump was near bankruptcy and Jewish bankers and financiers helped him out. His increasing the annual aid given to Israel was raised to five billion, recognition of Jerusalem as the capital of Israel, and his refusal to investigate 9/11 with the remark, "9/11, case closed" is questionable as to whether he is using his office to repay favors Jewish bankers and financiers have extended to him. Senators and Congresspersons have been guilty, but like the "Me Too" movement, it's time to put a halt to immorality.

# Bibliography

*Somdaily News,* An Australian publicaHe tion?

Catherine Beale, *"Phone Calls from Victims Trapped in*

*the Towers,"* Quora.

Cockburn, Andrew& Leslie, *Dangerous Liaisons,* 1991
Dan Edge, *The World*, *"Ft Carson Returning Combat
Veterans."*
*Veterans Today,* John Lear & Dimitri Khalezov articles

Greg Syzmanski, *William Rodriguez Testimony Destroys
9/11 Cover Up.*

*Five Dancing Israelis*, Armed Forces Radio and New
Jersey newspapers.

Michael Ruppert former head of LAPD Narcotics
Division  Author of the best seller, *Crossing the Rubicon*

Milan Simonich, *Pittsburgh Post-Gazette*, "Photographers
of 9/11"

William Rodriguez, North Tower janitor for 20 years.

Francesco Cossiga, from *Corriere della Sera.*

Jerry Mazza, free-lance journalist, New York City

Richard Gage, *Was 9/11 An Inside Job?*

Professor Steven Jones, *9/11 Demolition Using Thermite*
The Head of Flying School: *None could fly a Boeing 767*

1,270 Engineers say twin towers collapse was demolition.

Edward Snowden Reveals

Findley, Paul, *The Attack on the USS Liberty by Israel*, from The National Journal

Pilots@pilotsfor9/11truth.org.

*Forbes Financial Review*, November 28, 2011.

Gordon Duff, *Veterans Today* website.

Griffin, *David Ray, The New Pearl Harbor, Guns and Butter,* Professor Emeritus, Guelph University,

Barbara Honegger, *Behind the Smoke Curtain, Amazon.com*

Madsen, Wayne, *Washington Insider*, you tube, *Voices from the Grave.*

Marshal, Phillip, *The Big Bamboozle.*

Bowman, Robert, *There was no intercept by defense planes*

Deah M. Jackson, *Where Was NORAD on 9/11?*

Lindauer, Susan, *veteranstoday.com.author/susanlindauer*

Moret, Leuren, journalist, Israeli agents in Gaza fired rockets Into Israel to blame on Hamas.
Cynthia McKinney, Congresswoman who put her staff to find the truth about 9/11 and was unseated in her next election. She had courage while others didn't which may explain why few Americans have faith in either party or the government.

Hogdan, Robin, Logan Airport air traffic controller
Jones, Steven, PhD physicist and talk show host.
Judge, John, Aide to Congresswoman Cynthia McKinney
Khalezov, Dimitri, Russian S identifies 9/11 Mastermind
Nelson, George, Wittenberg, Russ, Captain, professional
    pilots for 9/11 Truth
Colin Alexander , *The Key*, a video on you tube, tells how
    **Flight 175** images were put on videos like AP's Kai
    Simonsen's video shot from a helicopter five miles away
from the South Tower which he was video-taping.
*The Hypocrites: A CIA/Mossad Road Map of Iran,*
by Andrew M. Patterson, available on Amazon com


*The Age of Disinformation,*  nine stories of the truth that tell
how wars are won and governments are run on lies.


*Deadly Diplomacy: How the UN Was Used to Destroy Bosnia,*
*by Richard Holbrooke, Bill Clinton, and Strobe Talbbot*


*Manipulation and Deception:* the hidden wars that killed
over 800,000 people and five and a half million homeless.


 Benjamin Freedman :*How President Wilson was manipulated*
into WW I.  Benjamin Freedman gave his entire wealth for peace.
Norman Finkelstein: *The Holocaust Industry*
Jack Cole videotaped the Auschwitz hospital that took care of
inmates and shows the gas chamber was an air raid shelter.
President George W. Bush refused to allow the International Red
Cross entry to Guantanamo or any other prison but Germany
under NAZI Party permitted the International Red Cross access to
any prison, concentration camp, without prior notice and talk to
any inmate at any time.

www.ingramcontent.com/pod-product-compliance
Lightning Source LLC
Chambersburg PA
CBHW070225190526
45169CB00001B/86